How to Solve Problems

Donald Scarl

How to Solve Problems

For Success in Freshman Physics, Engineering, and Beyond

Sixth Edition

Dosoris Press

Glen Cove, New York

HOW TO SOLVE PROBLEMS

Published by
DOSORIS PRESS
P. O. Box 148
Glen Cove, NY 11542
516-671-0686
http://www.dosoris.com
dosoris@dosoris.com

1st edition	1989
2nd edition	1990
3rd edition	1993
4th edition	1994
5th edition	1998
6th edition	2003

Library of Congress Control Number: 2003095140

ISBN: 0-96220086-7

Printed in the United States of America

9 8 7 6 5 4 3 2 1

Contents

4 Describing the Problem

5 Finding the Solution

6 Presenting the Results

7 Can't Solve It

8 Spreadsheets

Preface

To solve science and engineering problems you need to know science and engineering.

You also need a tool-kit of problem-solving methods.

This book describes the methods professional problem solvers use, explains why these methods have evolved, and shows how to make them your own. You can use these problem-solving methods now for homework and examinations. You will use them in your later professional work to help you define useful problems, solve them, and convince the world that the problems are important and that your solutions are correct.

Starting with the pioneering work of Polya, many authors have written self-help books and scholarly books on problem solving. These books are about the thinking part of problem solving: how to generalize, specialize, particularize, brainstorm, and so forth. They are especially useful for design problems in which part of the solution may lie in redefining the original problem.

This book, in contrast, emphasizes the simple actions that professional problem solvers take to analyze and solve already defined problems. It explains how to set up and solve problems that you did not think you knew how to solve. It describes the statement, organization, and presentation of scientific and technical material. By teaching problem-solving style, it attempts to do for problem solvers what Strunk and White's *Elements of Style* has done for writers.

It is a pleasure to acknowledge Eric Rogers, Jay Orear, and George Wolga, who by their instruction and example helped me learn how to teach. Hilda Bass, Barbara Cohen, Lorcan Folan, Robert Folk, Judith Scarl, Romualdas Sviedrys, Robert Swart, Larry Tankersley, and Alan Van Heuvelen read and improved manuscripts of this or previous editions. I have enjoyed a discussion with Annalisa Crannell about her methods of teaching mathematical writing.

Please send comments about this edition of *How to Solve Problems* to dosoris@dosoris.com.

1

Why Solve Problems?

How can I pass freshman physics? How can I build a user-friendly computer? How can I sell a million of them? How can I get a date for tomorrow night?

We all need to solve problems. This chapter explains how problem solving, in addition to helping you to pass science and engineering courses, can help you to learn technical material, to work efficiently, to think clearly, to develop self-confidence, and to convince others to work with you and to support your work.

Lifetime learning

Science, engineering, and life itself change from year to year, so that most of your learning will be done after you leave school and no longer have teachers, homework, tests, and grades to help you. Learn in school how to learn by yourself.

A good way to learn a new subject is to make up problems to solve for yourself. By making up problems you can find the gaps in the way you—and perhaps others—understand a subject. Problem solving is a path to new knowledge and discoveries.

You start out solving problems for others. Your teachers suggest that you do homework problems and require that you do exam problems. Your employer will ask you to solve problems that are part of a group engineering effort. But problem solving is most effective when you begin to solve problems that you make up yourself to satisfy your curiosity, to teach yourself a new subject, or to invent or discover something useful to society.

Even when doing problems for others, you will do them better and enjoy them more if you pretend that you are doing them for yourself. Doing problems as if for yourself is a good habit.

Workmanship

Human beings have always solved problems; it is part of what makes us human. Problem solving has led to progress, health, security, and comfort. But almost as significant as the solution of a problem is the workmanship used to get that solution. Workmanship, whether in carpentry, writing, or mathematical analysis, is something that can be learned. Part of the pleasure of problem solving is the sense of workmanship and self-esteem that it brings.

Previously solved problems

Although we like to think that the long-lasting things we learn in school and in life are *concepts*, we also come back again and again to a set of solved problems. When we want to solve a new problem, we try to make it similar to one of our already solved problems. Before an exam you will look at your already solved homework and quiz problems to review because there is a good chance the exam problems will be similar to them. As a professional scientist or engineer you will look at your solved and recorded problems to get ideas about how to solve new problems. If you document your solved problems with explicit assumptions, drawings, symbol definitions, complete algebra, and clearly stated results, you will still be able to use them a few weeks or many years later.

Reflexes

By developing good reflexes you can reduce the amount of thinking you have to do to solve problems. You will be able to use the same series of steps to begin every problem. When these steps are automatic, you do not have to think about them, so you can save your thinking for the harder parts of the problem. Just as in walking, driving a car, or playing tennis, as you make more actions automatic, you can achieve higher levels of performance.

The steps described in this book will become automatic, raising your ability and performance in problem solving. They apply not only to science and engineering problem solving but to all thinking. Without learning good reflexes you may be able to crawl, but you will not be able to win the race.

Presenting work to others

The homework, term papers, lab reports, and tests that you give to your teachers to evaluate are the public part of your work in school. As a professional engineer or scientist you will give informal talks to your research group, write quarterly research reports, give talks to technical societies, and write papers for technical journals. The way you analyze, solve, and present problems will help to determine your career success. While you are in school, homework and tests prepare you for these scientific reporting activities.

It is frightening to do work that others will see. "What if I'm wrong?" "What if they think that I'm not smart?" One of the ways to get over these feelings is to hand in work, year after year. The journalist who writes an article every day soon stops worrying about showing his work to others. When you hand in homework each week and take tests every few weeks, you will conquer the fear of having others evaluate your work.

The fear of showing work to others sometimes leads to lack of effort and carelessness. The argument goes something like this: "If I don't try too hard and am sloppy, it shows that I really could do better." This is a deadly argument and wrong. You will be judged only on the work that you *do*, not on the work that you *could* do. The right attitude is: "I am going to do my best on every piece of work I do, whether I show it to someone else or not."

Barriers

Thinking is difficult. Mental barriers do not allow young unformed thoughts to pass from the back of your mind to the front. Identify and eliminate your mental barriers.

"I don't think I can do this problem." There is nothing that can stop you as fast as that thought. Do not put that unnecessary barrier in your way. Think "I will be able to do this problem, although it may be harder than I thought and may take more work than I thought."

"I'm afraid I will make a mistake." Fear of mistakes can stop progress. Learn how to find and correct mistakes, but have no fear of them. Everyone makes mistakes. No one goes from the beginning of a problem to the end correctly the first time. (If you do, you are not doing hard enough problems.)

"I can't do mathematics (physics, French, ...)" Those who seem to be able to do things more easily than you almost always have learned better methods, have developed better work habits, or are simply working harder. In college talent plays only a tiny role in our ability to learn and succeed. It may be true that talent separates those at the top of their professions from others, but it makes only a minor difference in school.

"This is boring; I would rather be surfing." It is easy not to be interested in the work you have to do. Surprisingly, it is also easy *to* be interested in the work you have to do. Interests are not in us from birth. They change. They can be changed. A powerful teacher, a clear book, or a television program all can give you a new interest in a subject. You can also do it yourself. "I need to do this, so I *am* interested in it and will try to figure out how to do it well." Wide and enthusiastic interests improve your life.

You can increase your interest in science, your ability in science, and your self-confidence by learning to solve problems smoothly and well.

Creativity and disorganization

Although the procedures presented in this book are methodical, rational, and powerful, a small number of creative people achieve extraordinary results without them. These people are intuitive, impulsive, and impatient. They have difficulty organizing and explaining their thoughts, their work, and their lives. Some good engineers and scientists fall into this category. Methodical procedures slow them down. That is fine for them, but not for you. The great majority of people, and certainly of scientists and engineers, work far better and more productively using methodical procedures. Be one of them.

2

School

In school you learn concepts, facts, and procedures and develop good intellectual, professional, and social habits. This book teaches procedures and habits rather than concepts or facts. These procedures and habits will help you to learn concepts and facts, to interact with others, and to get the pleasures and rewards of scientific and technical accomplishment.

Science and engineering texts

"I understand the material; I just can't do the problems." *Understand* is a slippery word. If the first time you read something it seems to make sense, it is easy to say, "I understand that." Making sense is not enough. Understanding means getting exact knowledge of each part of an idea and knowing how to use that idea in new ways that were not included in its explanation.

Science and engineering books are written differently from other books. While in other books there is sometimes one new idea per paragraph or per chapter (or sometimes per book), in science and engineering books there is often one new idea per sentence. It is not surprising that it takes longer and requires more effort to read these books than others. They are designed not just to be read but to be worked on with pen and paper. When you switch from reading literature to reading physics, it is sometimes hard to remember that you need to write to think. Work out the mathematical steps that are left out in each explanation and make up and work out problems of your own that lead to real understanding.

You can understand each new idea by doing problems using that idea. What if there are no problems? Make up your own. Develop this habit that allows you to learn and understand on your own. Read an idea. Turn it over in your mind. Ask yourself questions. Make up a simple problem. Invent a problem testing an extreme case. Solve the problems.

Since no one grades the problems you make up and solve yourself, they can be fun, so much fun that some of the best scientists and engineers would rather solve problems than do anything else.

Learning equations

One aspect of problem solving works backwards. By analyzing the kind of information you need to solve problems, you will be able to ask yourself the right questions when you are learning new material.

"I don't know what equations to use to solve this problem." You can prepare yourself for choosing the right equations by classifying and describing each equation as you learn it. Give the equation a name, describe its importance, classify it, write down the conditions under which it holds, and write a list of the definitions of the symbols used in the equation. Describing equations as you learn them is the key to choosing the right equations to solve a problem later.

An equation's importance

Not all teachers are willing to admit it, but some equations are much more important than others. In introductory physics $F=ma$ takes the prize, with $K=\frac{1}{2}mv^2$ and $p=mv$ close behind. An equation can be important because it is a definition, because it is true under many conditions, because it is simple, because it cannot easily be derived, because it can solve a lot of problems, because it is famous, or because your teacher likes it. While you are learning an equation in a course, its importance goes up if your teacher mentions it in lecture or assigns a homework problem or a quiz problem using it. Learn to be a connoisseur of equations.

Classifying equations

Classify each equation as a definition, a fundamental law, or the solution of a particular problem. If an equation is a definition or a fundamental law, you probably need to memorize it. If an equation is the solution of a particular problem, maybe you can derive it easily, or maybe the problem is specialized enough that you can look up the equation whenever you need it and do not have to memorize it.

Do not spend your time memorizing every equation you read. Go through life knowing the smallest possible number of equations. Spend your time learning the conditions under which each important equation holds and the exact meaning of each symbol in that equation. It is better to use a few tools well than many badly.

For each equation ask yourself: What kind of problem would this equation help me solve? Is it practical to turn this equation around so that I can find one of the quantities on the right hand side in terms of the quantity on the left hand side? To find one of the quantities, what other quantities do I need to know?

Special conditions for each equation

An equation holds only under certain conditions. Some equations are much more limited than others. While $F=ma$ is always true, $x=x_0+v_0t+\frac{1}{2}at^2$ is very seldom true. (It is true only for motion with constant acceleration.) Many of the mistakes made in doing problems come from using an equation that doesn't hold under the conditions of the problem. Write down the special conditions when you learn the equation.

Definitions of symbols

Know the exact definition of each quantity that enters an equation. If g is the earth's gravitational constant at the surface of the earth, it will not be the earth's gravitational constant at a satellite that is 20,000 km above the surface. Write down the exact definitions when you learn the equation.

Write specific and exact definitions. "The horizontal distance from the starting point" is better than "the distance." "The angle of the initial velocity above the horizontal" is better than "the angle."

Example: How to classify equations

As an example, here is a selection of equations from one chapter of an excellent and popular introductory physics text.

$$v = \frac{d\mathbf{r}}{dt}$$

$$x-x_0 = (v_0 \cos\theta_0)t$$

$$y = (\tan\theta_0)x - \frac{g}{2(v_0\cos\theta_0)^2}x^2$$

$$a = \frac{v^2}{r}$$

$$v_{PA} = \frac{v_{PB} + v_{BA}}{1+v_{PB}v_{BA}/c^2}$$

After reading a section in the text that presents one of these equations, describe and classify the equation on a sheet of paper. When you are finished with the chapter, your sheet might look like this.

▽ ▽ ▽ ▽ ▽ ▽ ▽ ▽ ▽ ▽ ▽ ▽ ▽

Definition of velocity:	$\mathbf{v} = \frac{d\mathbf{r}}{dt}$

Importance: Basic, memorize it

Class: Definition

Conditions:	Always holds	
Symbol definitions:	vector velocity of a particle	$= \mathbf{v}$
	radius vector from origin to particle	$= \mathbf{r}$

Component equations:

$$v_x = \frac{dx}{dt}$$

$$v_y = \frac{dy}{dt}$$

$$v_z = \frac{dz}{dt}$$

Horizontal position of a projectile: $x - x_0 = (v_0 \cos\theta_0)t$

Importance:	Not very. I can derive it when I need it from the equations for motion under constant acceleration. Was not mentioned in lecture.	
Class:	Solution of particular problem.	
Conditions:	Only for horizontal motion of a free particle with no air resistance.	
Symbol definitions:	horizontal position at t=0	$= x_0$
	horizontal position at time t	$= x$
	speed at t=0	$= v_0$
	angle of velocity vector above horizontal at t=0	$= \theta_0$
	time at which x is measured	$= t$

Trajectory of a projectile: $y = (\tan\theta_0)x - \dfrac{g}{2(v_0\cos\theta_0)^2}\, x^2$

Importance:	Little. It shows that the trajectory is a parabola. Very complicated and I can derive it from the equations for constant acceleration. Was not mentioned in lecture.
Class:	Solution of particular problem.
Conditions:	Vertical position of a particle as a function of its horizontal position. For a particle moving under gravity alone. No air resistance.

Symbol definitions:	vertical position	$= y$
	horizontal position	$= x$
	size of initial velocity	$= v_0$
	angle of initial velocity vector above horizontal	$= \theta_0$
	acceleration of gravity at earth's surface	$= g$

Motion in a circle: $$a = \frac{v^2}{r}$$

Importance:	Important. Memorize. Use dimensions to check.	
Class:	Special problem, but very general.	
Conditions:	For motion in a circle with constant speed, but is a good approximation to any motion along a curved path.	
Symbol definitions:	acceleration toward center	$= a$
	tangential velocity	$= v$
	radius of circle	$= r$

Relativistic addition of velocities: $$v_{PA} = \frac{v_{PB} + v_{BA}}{1 + v_{PB}v_{BA}/c^2}$$

Importance:	Advanced. Look it up when I need it.
Class:	Special problem. Relative motion of one particle with respect to another when the speeds are close to the speed of light.
Conditions:	Two particles moving along the same line. Works for any velocity but is only necessary when at least one of the velocities is near the velocity of light.

Symbol definitions:	velocity of particle A with respect to my coordinate system	$= v_{PA}$
	velocity of particle B with respect to my coordinate system	$= v_{PB}$
	velocity of particle A with respect to particle B	$= v_{BA}$
	speed of light	$= c$

△ △ △ △ △ △ △ △ △ △ △ △ △

Your sheets describing the equations you are learning, the problems you have made up and worked using these equations, and your thoughts about the meaning of what you have done are the first steps on the road to understanding.

Homework

The way you solve homework problems will grow into the way you solve engineering problems. Since school is for practice, your first work in a course does not have to be perfect. Your ability will increase as the semester proceeds. Sometimes you won't notice the improvement, since the material also gets harder as the semester proceeds. Notice, though, that last week's homework seems easy to do this week.

If you do homework problems in a way that is easy for your teacher to understand, they will also be easy for you to understand when you use them to study for the next exam. Your homework and exam grades will improve and your teacher's opinion of you will improve too. The skill you develop doing homework problems carefully will carry over into exams, improving your grades, and into your professional life, increasing your ability and satisfaction.

Exams

Which parts of an exam are the most important? The answers. Yes, but... By concentrating on the answers during the whole exam, your answers will get worse, not better. If you read the problem quickly and interpret it wrong, your carefully worked out answer will be wrong. If you work quickly and make a mistake in algebra, your answer will be wrong. If you work quickly and do not write down all your steps, the person who corrects the exam cannot tell what you were doing or how much you know and cannot give you the partial credit you deserve.

The methods suggested in this book seem at first to be impossible to use on exams. How can you work carefully and slowly when there is only one hour to do six problems or even twenty problems? You can. You must. Doing homework problems every week using the proper methods will give you enough practice that you can use the same methods on exams, even when you have less time than you need. These methods will increase your probability of getting right answers and increase your grade. They will increase the amount of partial credit you get when your answer is wrong. They will show your teacher that you understand the material of the course and can work in a professional way.

Sometimes exams look different from other calculations. In particular, multiple-choice problems are dangerous. A multiple-choice problem that has several answers already written down still requires the same calculation on paper as any other problem. If you try to do a multiple-choice problem in your head, you are regressing to old problem-solving habits and decreasing your probability of getting the right answer. Use a sheet of paper or a page of the exam booklet to work out each multiple-choice problem. Find the answer and rejoice when it agrees with one of the choices. Be careful about mistakes, because some of the multiple choice answers will purposely be the ones generated by common mistakes.

You can fall into the same trap in exams as in homework if you work quickly and sloppily with the subconscious idea that "This is not my best work, I really could do better if I wanted to." Force yourself to do exactly your best work. When that best work is not good enough, ask, "How can I do better next time?"

Efficiency

"Doing problems in my head or just putting down a few steps and getting the answer is quick and efficient." "I don't have enough time to work carefully on exams." "I got all A's in high school but I can't pass college physics." These three thoughts are connected. The method that is most efficient for simple problems is no longer the most efficient for harder problems. One can not expect the method that works well for "If the distance from Lima, Peru to Timbuktu is 3600 miles and an airplane goes 360 miles per hour ..." to continue to work well for "Find the optimum wing design for maximum lift, minimum drag, minimum weight, minimum moment of inertia, and maximum strength, using graphite fiber composite material." The methods described in this book have been developed by professional engineers, scientists, and designers because they are efficient in getting the right answer, in sharing work with colleagues, in recalling and modifying past work, and in reporting work to others. These methods are efficient for homework, exams, and research calculations.

Teachers

After teaching a course several times, your teacher will know more about it than you do. There are several ways you can react to this: "There is so much to know, I will never learn anything," is one wrong way. "If she has been able to understand, I can too," is one right way. Remember that it has taken your teacher a long time to learn what she has learned and that some of what she knows she has learned by talking with previous students and answering their questions. On the other side, your teacher, if she has continued to work in her field, realizes that she does not yet know everything there is to know about her subject. She is learning too, knows how hard it is to learn, and is glad to accept all the help she can get.

A course is a two-way contract. Your teacher influences you by the work she assigns, by the tests she gives, by her explanations, and by her attitude toward you. You influence your teacher by your attitude toward the material and toward her. You will inspire her to teach better if you are interested in the material, ask questions during class and afterward, and talk to her during office hours at least twice each semester. Your interest will encourage her to continue to put effort into her presentation, find interesting examples, work on improving her explanations, and write clear, fair exam questions. Show your enthusiasm and help both your teacher and yourself to learn.

Going to a teacher during the semester is always a good idea even if the teacher is busy or is difficult to understand. Most teachers will be glad to see you a few times a semester. You can go just to say, "I enjoyed the example you did this morning," or "Where can I find more about synchronous orbits?" You can go to ask a question about what you are learning. Even if you display ignorance about the material in the course, you will be showing your interest in it. Asking relevant and precise questions is another part of scientific and engineering work that you can practice while you are in school.

You can also go to complain. Sometimes teachers don't know the difficulties students are having with the course or with assigned problems. Asking a teacher a question about a problem can give her useful information and help everyone in the course. If you and other students are having trouble with a course, first tell the teacher. Often, things will improve.

Working in a group

Most modern research and design is done in groups. The ability to solve problems as part of a team is not automatic; it has to be learned. Working with others on homework problems is one way to learn problem-solving teamwork. Each member of the team contributes ideas, each idea is considered and tried, and the best survive. The way that a good team can solve a problem that none of its individual members could solve sometimes appears magical.

In a good team each idea is accepted and appreciated; it is written down, its consequences are explored, it is connected to other ideas, and it is allowed to generate new ideas. When the problem is finally solved, each member of the team understands each of the ideas that went into the solution, whether contributed by herself or by someone else.

However, teamwork has dangers. If the group does not accept and encourage every member's ideas the most aggressive members will work on their own—not necessarily correct—ideas and the others will sit quietly and listen. The talkers will often go down the wrong path and the listeners will follow them. Or, if the talkers do find the right path, the listeners will accept their ideas, copy down the steps and the result, and believe that they understand how to solve the problem. When a similar problem appears on an exam, where no cooperation is allowed, the listeners will often find that their understanding was not as complete as they thought and that their grade is not as high it could have been. It is better to work individually than in a group of this kind.

If you cannot solve the homework problems that you need to solve, joining a group can help. Now your responsibility is to make sure that you question each step the group takes until you understand everything that the group does. This is socially difficult. Even in a group made up of two people, it is hard to say, "I don't understand that." It is hard on the ego and it seems to slow down the work. Only the best and most confident scientists and engineers are able to say it easily. In fact, one person saying "I don't understand that," often makes others realize that they don't understand either, and the resulting explanation helps the whole group. "I don't understand that" is one of the most valuable comments in group work.

Professional qualities

Can you develop, while in college, the qualities that will make your professional life more rewarding? Yes, if you know what those qualities are and work toward possessing them. Problem solving can help you reach that goal.

One professional quality is getting the right answer. Logical problem-solving methods will help you to get the right answer on tests in school and on professional calculations.

Knowing where to find needed information is a quality that is hard to practice in school. Teachers too often give you everything that is needed. Try to practice finding new information and become familiar with the standard sources of technical information.

Modifying and using past work is a good way to increase your professional efficiency. Doing calculations in a way that lets you come back to them a year later and understand what you did is difficult but valuable.

Sharing your work with others in your group and eventually leading a group can be practiced in college. Learn how to work with a group and how to write up your work so that others in the group can understand it.

Be optimistic. Confidence that you can solve a problem allows you to do the parts that you *can* do while not worrying about the parts that you *can't* do. If you believe that the job can be done and that you can do it, you will be able to convince others to support you and to work with you in a group.

Convincing your supervisors or the scientific community of the worth of what you plan to do is one of the skills that will determine your career success. Both new and continuing projects need the support of others. You must have your plans and calculations clearly written out, and therefore clear in your mind, before you can explain them in a convincing way to others, justify them, and defend them. Presenting a problem and your results to your supervisors or to the scientific community is an essential part of technical work. Practice while in school by stating each problem and its solution clearly on paper.

Exercises

2–1. Make a list of all of the major equations in Chapter 2 of your physics textbook. Write the equation, the conditions under which it holds, the definition of each symbol, the class of the equation, and the importance of the equation.

2–2. One of the most famous equations in the world is

$$PV = nRT\,.$$

Classify this equation, describe the conditions under which it is true, and write a careful definition of each of its symbols. Show units for each of the symbols.

2–3. A surprisingly important equation that comes from three-dimensional geometry, but is used in linear algebra and elsewhere, is

$$d^2 = x^2 + y^2 + z^2\,.$$

Classify this equation, describe the conditions under which it is true, and write a careful definition of each of its symbols. Draw a diagram to help in understanding the symbols. If you do not know what this equation is for, ask a classmate, but do not let them solve the problem for you.

3

Methods

This chapter describes methods that make problem solving easier, that allow clear thinking, and that lead to correct answers. These methods are essential when you need to solve problems that you do not know how to solve at first glance. They apply to all parts of the solution of a problem.

The next sections are an example of a problem and its solution. The solution has been done using the standard problem-solving methods that we will discuss in this chapter. When we introduce these methods, we will use parts of the solution as illustrations.

▽ ▽ ▽ ▽ ▽ ▽ ▽ ▽ ▽ ▽ ▽ ▽ ▽

Example1: Sears Tower Elevator

The Skydeck in the Sears Tower in Chicago is on the 103rd floor. The Skydeck elevator leaves the lobby and accelerates upward with a constant acceleration of 0.80 m/s^2 until it reaches its maximum velocity of 18.2 miles per hour. Then it travels at its maximum velocity until it passes the 94th floor, 1263 feet above the lobby.

a. How long does the elevator take to accelerate to its maximum velocity?

b. How high above the lobby is the elevator just as it gets to its maximum velocity?

c. How many seconds after it leaves the lobby does the elevator pass the 94th floor?

Solution for Example 1

Sears Tower Elevator

D. Scarl

12 July 2003

DRAWING

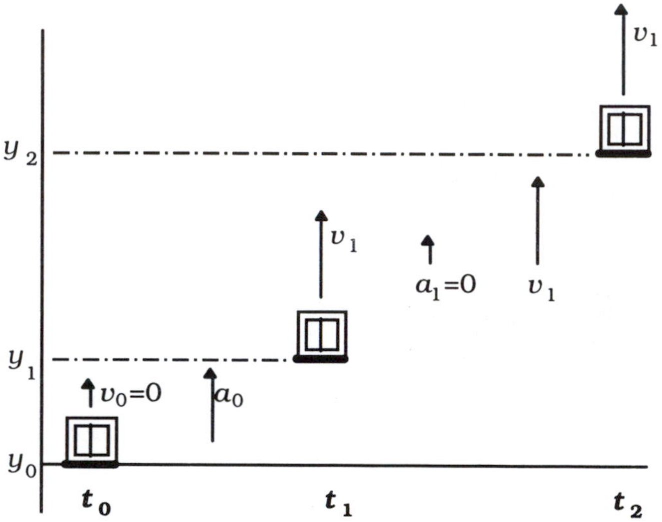

DEFINITIONS

At the lobby		
time	$= t_0$	$= 0$ s
height	$= y_0$	$= 0$ m
velocity	$= v_0$	$= 0$ m/s
From lobby until elevator reaches maximum velocity		
acceleration	$= a_0$	$= 0.80$ m/s^2
At the time that elevator reaches maximum velocity		
time	$= t_1$	
height above street	$= y_1$	
While elevator is traveling at maximum velocity		
acceleration	$= a_1$	$= 0$ m/s^2
velocity	$= v_1$	$= 18.2$ mi/hr
When elevator passes 94th floor		
time	$= t_2$	
height above street	$= y_2$	$=1263$ ft

CONVERT UNITS

$$v_1 \quad = 18.2\,\frac{\text{miles}}{\text{hour}}\left(\frac{1609\text{ km}}{1\text{ mile}}\right)\left(\frac{1\text{ hour}}{3600\text{ s}}\right) = 8.13\text{ m/s}$$

$$y_2 = 1263 \text{ feet}\left(0.305\,\frac{\text{meter}}{\text{foot}}\right) = 385 \text{ meters}$$

---

a) Find time to reach maximum velocity
General equations:
For motion with constant acceleration

$$y = y_0 + v_0 t + \tfrac{1}{2} a t^2$$

$$v = v_0 + at$$

When elevator reaches max velocity

$$v_1 = v_0 + a_0 t_1$$

$$v_1 = 0 + a_0 t_1$$

$$t_1 = \frac{v_1}{a_0}$$

$$= \frac{8.13 \text{ m/s}}{0.80 \text{ m/s}^2}$$

$$t_1 = 10.16 \text{ s}$$

The time for the elevator to reach its maximum velocity is $t_1 = 10.16$ s.

---

b) Find height of elevator when it reaches its maximum velocity.

$$y = y_0 + v_0 t + \tfrac{1}{2} a t^2$$

$$y_1 = 0 + 0\, t_1 + \tfrac{1}{2} a_1 t_1^{\,2}$$

$$= \tfrac{1}{2}(0.80 \text{ m/s}^2)\,(10.16 \text{ s})^2 = 41.29 \text{ m}$$

When it reaches maximum velocity, the elevator is 41.29 m above the lobby.

---

d) Calculate time at which elevator passes 94th floor.
While the elevator is moving with its maximum velocity

$$y = y_0 + v_0 t + \tfrac{1}{2} a t^2$$

$$y_2 = y_1 + v_1(t_2 - t_1) + 0$$

(The elapsed time while the elevator is moving at its maximum velocity is $t_2 - t_1$.)

$$t_2 - t_1 = \frac{y_2 - y_1}{v_1}$$

$$t_2 = t_1 + \frac{y_2 - y_1}{v_1}$$

$$= 10.16\ \text{s} + \frac{385\ \text{m} - 41.29\ \text{m}}{8.13\ \text{m/s}}$$

$$= 10.16\ \text{s} + 42.30\ \text{s}$$

$$t_2 = 52.46\ \text{s}$$

The elevator moves with its maximum velocity for 42.3 s and passes the 95th floor 52.5 s after it leaves the lobby.

△ △ △ △ △ △ △ △ △ △ △ △ △

Divide into parts

Dividing the solution into parts is the first step in solving any problem. It is useful for problems at all levels and is essential for large problems. (Professional programmers know that one of the hardest tasks in writing a large program is dividing the program into parts that can be worked on separately and recombined smoothly.) When starting a problem, divide the description of the problem and its solution into the smallest possible parts and work on each part separately.

The parts into which most introductory science and engineering problem solutions can be divided are

Heading
Labeled drawings
Symbol definitions
Data
Preliminary equations
Science equations
Calculation
Results.

The first four of these parts are a restatement of the problem. They will be described in Chapter 4. The next three, the calculation of the solution, will be described in Chapter 5. The last part, presentation of the results, will be described in Chapter 6.

After you have divided the problem into parts, divide each of the parts into parts. One way you can do this is by dividing each part in space and time. If there are several different masses, describe the conditions for each mass separately. If there is a circuit with many amplifiers, first draw and analyze each amplifier separately. If the events in the problem take place at two or more different times, describe separately what is happening at the first time, the second time, and so forth, then write the equations that connect the events at the different times.

Do the parts separately

First divide a problem into parts; then force yourself to think about the parts separately. Do not try to solve the whole problem at once. As you start to work on a problem, it is tempting to try to figure out the answer or to worry about what equations apply and how to solve them. These are not the parts of the problem to think about at the beginning. First do the simple and automatic steps that begin the solution of every problem: write a heading, draw a labeled diagram, define your symbols. None of these steps requires knowledge of how you will eventually solve the problem.

Professional problem solvers learn not to worry about the parts of the solution that they cannot do, while they work on the parts they can do. If a problem is difficult, describing it clearly and dividing it into parts is a useful contribution, even if some of the parts cannot be done by you or by anyone else. Describing what needs to be done is a first step toward getting it done.

Dividing a solution into parts is useful on tests. If you sit and worry about the part of the problem you can't do, you will get no credit for writing down the parts you can do. After writing all of the automatic steps, such as the definition of symbols and the geometry equations, you may understand the problem well enough to be able to solve it. Divide the solution into parts and do the parts you can do.

Label each part

Begin each part of the solution with a short title saying what you are starting to do.

▽ ▽ ▽ ▽ ▽ ▽ ▽ ▽ ▽ ▽ ▽ ▽ ▽

Drawing

⋮

Definitions

⋮

Unit conversion

⋮

Geometry

⋮

General equations

⋮

△ △ △ △ △ △ △ △ △ △ △ △ △

Writing a title for each part of the solution helps in thinking about that part of the solution and helps even more when you come back to the problem later or show the problem to someone else. On tests, writing titles helps your instructor understand what you are doing and leads to a higher grade. These titles are part of the documentation of the problem; they make clear what you are doing on each line.

A special case of labeling each part of the solution is writing a one-line description of the problem at the beginning of your work.

▽ ▽ ▽ ▽ ▽ ▽ ▽ ▽ ▽ ▽ ▽ ▽ ▽

Sears Tower Elevator

⋮

△ △ △ △ △ △ △ △ △ △ △ △ △

This title helps those absent-minded professors who do not remember exactly what problem they assigned and helps *you* remember exactly what the problem was when, before an exam, you look back at a problem that you solved at the very beginning of the course.

Algebra is a very condensed way of expressing ideas. Break up the algebra with a few words describing what you are about to do.

▽ ▽ ▽ ▽ ▽ ▽ ▽ ▽ ▽ ▽ ▽ ▽ ▽

⋮

Find time to reach maximum velocity

⋮

△ △ △ △ △ △ △ △ △ △ △ △ △

Finally, some labels are just English sentences saying what your result was.

The time for the elevator to reach its maximum velocity is $t_1 = 10.16$ s.

This sentence makes it easier to understand your result, helps you to think about whether it makes sense, and gives your result its proper importance.

Work down the page

The graphic arrangement of your work can make it easier to read. Work down the page, one thing under the next. Jumping to a second column or writing things side by side interferes with being able to understand the order of your solution. Instead of writing up the side of the page or squeezing in the last few lines at the end of a page, start a new numbered page. Think of the person reading or grading your paper. Can she understand the order of your solution and give you credit for all the parts that you have done?

The right hand side of an equation can go across the page with successive equal signs but it is better to start a new line for each new equal sign. When you go to the next line with another step of the same equation, align the equal sign with the one above.

▽ ▽ ▽ ▽ ▽ ▽ ▽ ▽ ▽ ▽ ▽ ▽ ▽

When elevator reaches max vel

$$v_1 = v_0 + a_0 t_1$$

$$v_1 = 0 + a_0 t_1$$

$$t_1 = \frac{v_1}{a_0}$$

$$= \frac{8.13 \text{ m/s}}{0.80 \text{ m/s}^2}$$

$$t_1 = 10.16 \text{ s}$$

△ △ △ △ △ △ △ △ △ △ △ △ △

Write on one side of the paper. Compared with the cost of your education, paper is inexpensive. When you are doing a long calculation, it is annoying to have to turn over pages to find the part of the solution you are looking for. The only exception to this is in test booklets or notebooks where writing on both sides is necessary and expected.

Use horizontal lines to separate parts of the solution or to show that you have stopped working on one equation and have begun working on another.

Use solid lines, dashed lines, dotted lines, squiggly lines, double lines, or whatever you like to show what is going on.

Use them where we use **boldface** in this book or wherever you want.

> Use boxes for answers. Finish each section of the problem by writing a short English sentence describing your result. Put the name of the variable and its unit in the sentence.

The more aids to the eye you use, the easier it is to understand your work. Make it easy for the person reading your solution (it could be yourself) to find what they need.

Write clearly

Your handwriting reveals your attitude. Slow down and write beautifully. Here again, teachers can set a bad example. They write quickly on the blackboard so that they can cover more material in an hour. When you are solving a problem, you do not have to write quickly. The time you spend writing is a small fraction of the time you spend working on the problem.

If the way you always write a symbol allows it be confused with another symbol, change the way you write that symbol. If you accidentally write a symbol or number unclearly, cross it out and write it again clearly. Write a zero before naked decimal points to make sure the decimal point doesn't get overlooked: $x = 0.17$ m. You will write each symbol in a long calculation many times. If it is unclear even once, it can lead to confusion and mistakes. You are the one who will suffer the most from writing that is not carefully done. Writing sloppily is like shooting yourself in the foot. It will slow you down later.

Make everything explicit

Write everything down. So easy to say, so hard to do. Facts that you keep in your head instead of writing them down make it much harder to think about the problem. Facts that you do not write down cannot be checked for mistakes. Facts that you do not write down cannot be used or even understood when you come back to the problem in a few years. Facts that you do not write down will generate no credit on an exam. If something is part of the problem, write it down.

▽ ▽ ▽ ▽ ▽ ▽ ▽ ▽ ▽ ▽ ▽ ▽ ▽

Definitions

At the lobby

time	$= t_0$	= 0 s
height	$= y_0$	= 0 m
velocity	$= v_0$	= 0 m/s

From lobby until elevator reaches maximum velocity

acceleration	$= a_0$	= 0.80 m/s2

△ △ △ △ △ △ △ △ △ △ △ △ △

Use symbols

At the beginning of your description of a problem make up a definition and a symbol for every quantity in the problem. Although the statement of the problem will usually contain numbers and your answer will usually be a number, you need symbols to describe the problem and to solve it.

Using symbols for quantities is hard to do at first. With symbols the equations look more complicated than they do with numbers. With symbols the equations look like algebra, while with numbers they look like arithmetic. Algebra works at least as well as arithmetic if you write every step and do one step at a time.

First set up and solve the problem with symbols. Use algebra to get the symbol for each answer by itself on the left hand side of an equation. Then replace the symbols on the right hand side of the equation by numbers to calculate the answer.

By working with symbols you can solve the problem in general. If the input numbers change, as they often do in a professional problem, all you have to do is put the new numbers in the algebraic equation for the answer and get a new result.

Often you will need to present the result of a professional problem as a graph of how the result depends on one or more of the initial conditions. You can calculate the graph easily if you have the answer as an algebraic expression. Just substitute different values of the initial conditions and calculate corresponding values of the result.

If you use numbers during the solution of a problem, you have to go through the whole solution changing numbers when the initial conditions change. Programmers know not to use numbers in the body of a computer program or spreadsheet because numbers are hard to find once they are absorbed into a calculation. Numbers belong only at the beginning,

▽ ▽ ▽ ▽ ▽ ▽ ▽ ▽ ▽ ▽ ▽ ▽ ▽

Definitions

At the lobby

time	$= t_0$	$= 0$ s
height	$= y_0$	$= 0$ m
velocity	$= v_0$	$= 0$ m/s

From lobby until elevator reaches maximum velocity

acceleration	$= a_0$	$= 0.80\ \text{m/s}^2$

⋮

△ △ △ △ △ △ △ △ △ △ △ △ △

and at the end.

▽ ▽ ▽ ▽ ▽ ▽ ▽ ▽ ▽ ▽ ▽ ▽ ▽

Find height of elevator when it reaches max vel.

$$y = y_0 + v_0 t + \tfrac{1}{2} a t^2$$

$$y_1 = 0 + 0\, t_1 + \tfrac{1}{2} a_1 {t_1}^2$$

$$= \tfrac{1}{2}(0.80\ \text{m/s}^2)\ (10.16\ \text{s})^2 = 41.29\ \text{m}$$

△ △ △ △ △ △ △ △ △ △ △ △ △

Another advantage to calculating with symbols is that you can see how the result depends on the initial conditions.

▽ ▽ ▽ ▽ ▽ ▽ ▽ ▽ ▽ ▽ ▽ ▽ ▽

When elevator reaches max vel

$$v_1 = v_0 + a_0 t_1$$

$$v_1 = 0 + a_0 t_1$$

$$t_1 = \frac{v_1}{a_0}$$

△ △ △ △ △ △ △ △ △ △ △ △ △

Does this equation for the time to reach maximum velocity show that the time increases as the maximum velocity increases? Does it show that the time decreases as the acceleration increases? These questions are good checks that you did the algebra correctly. They also let you understand the form of the result and its sensitivity to each of the initial conditions. Some of the quantities may even cancel, telling you that they were not essential parts of the problem and saving you time and errors in the final calculation with numbers.

By working out the solution with symbols you can check the units in the result, which is a good way to catch algebraic errors. If the unit on the right hand side of the equation for the result does not agree with that on the left, there is a mistake somewhere.

On the other hand, if, after choosing and defining symbols for a problem, you cannot see how to solve it, putting in numbers can sometimes help you to get started or help you to solve equations. Chapter 7 explains how this works.

Zero

There is an exception to the rule about saving numbers for the end of the problem. After you have written one of the full famous general equations that will help you solve a problem and have written a particular case of that equation using the variables you have defined for your problem, you can substitute zero for quantities that are zero.

▽ ▽ ▽ ▽ ▽ ▽ ▽ ▽ ▽ ▽ ▽ ▽ ▽

---

b) Find height of elevator when it reaches its maximum velocity.

$$y = y_0 + v_0 t + \tfrac{1}{2} a t^2$$

$$y_1 = 0 + 0\, t_1 + \tfrac{1}{2} a_1 {t_1}^2$$

$$= \tfrac{1}{2}(0.80\ \text{m/s}^2)\,(10.16\ \text{s})^2 = 41.29\ \text{m}$$

△ △ △ △ △ △ △ △ △ △ △ △ △

By putting in zeros as soon as possible (never in the original writing down of the famous equations) you can shorten the problem and solve it more easily. The price that you pay for this is, if the initial conditions change so that a variable is no longer zero, you need to go back and redo the entire solution. Even if this happens, you can use your first simplified version to check the more complicated result.

Do one step at a time

Do algebra one step at a time.

When doing algebra, write down every step, one step at a time. The step that you do not write down is the one that contains the mistake.

There are, unfortunately, too many reasons why writing one step at a time does not come naturally. Almost all teachers who write equations on the blackboard do several steps at a time. All textbooks and technical papers do several steps at a time. They do this to save time and paper. You can be sure that these same teachers and authors worked out the solution one step at a time before they wrote on the blackboard or in their book. The way a solution is finally presented in a lecture or book is not a good model to use when doing your own problem solving.

The reason for doing one step at a time is to avoid mistakes. If each step is not written down separately, each step cannot be checked for mistakes. A single algebraic mistake makes all of the results that follow it wrong. Mistakes near the top of a solution lead to even more trouble than those near the bottom. Mistakes are expensive. They waste time and generate frustration. Work in a way that minimizes mistakes.

One step at a time.

Use ratios

A ratio (one quantity divided by another) can often be found even when the quantities themselves cannot. Sometimes the data in a problem are ratios. Sometimes you can most easily calculate a result as the ratio of an unknown quantity to a known quantity.

In problems in which the data contain a ratio it is often easy to calculate the result as a ratio.

▽ ▽ ▽ ▽ ▽ ▽ ▽ ▽ ▽ ▽ ▽ ▽ ▽

Example 2: Surface area of a sphere

If the diameter of a sphere is doubled, by what factor does the surface area of the sphere increase?

Solution for Example 2

Surface area of a sphere

Name
Date

Diagram

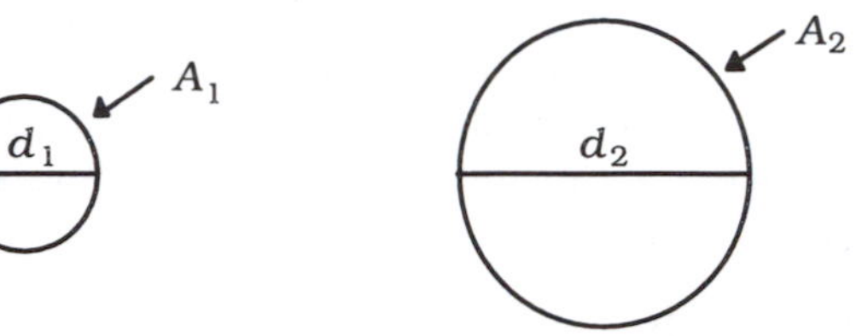

Definitions

diameter of original sphere	$= d_1$		m
diameter of new sphere	$= d_2$	$= 2d_1$	m
surface area of original sphere	$= A_1$		m^2
surface area of new sphere	$= A_2$		m^2

Equation for the surface area of a sphere

$$A = 4\pi r^2$$

In terms of diameter

$$A = 4\pi r^2 = 4\pi(\frac{d}{2})^2 = \pi d^2$$

Ratio of new to original surface area

$$\frac{A_2}{A_1} = \frac{\pi d_2{}^2}{\pi d_1{}^2} = \left(\frac{d_2}{d_1}\right)^2 = 2^2 = 4$$

$$A_2 = 4A_1$$

> The surface area of a sphere whose diameter was doubled is 4 times its original area.

△ △ △ △ △ △ △ △ △ △ △ △ △

The ratio of two quantities, each of which have the same units, is dimensionless and has no units. A dimensionless ratio gives the value of the top quantity measured in units of the bottom quantity. For instance, an acceleration of a=19.6 m/s^2 can be written a=2g since a/g =19.6 m/s^2 / 9.8 m/s^2 = 2. A dimensionless ratio also tells what fraction the top quantity is of the bottom quantity.

There are many famous and useful dimensionless ratios that describe physical conditions, like the Reynolds number in fluid dynamics and the Strehl ratio in optics.

People have discovered new physical laws by noticing that an experimental result depends only on the ratio of two quantities or that the ratio of two quantities is always a constant.

Do the assigned problem

In a homework or test problem make sure to do the problem that is written down and to find the solution that is asked for. Do not do more or less than the problem asks. If the instructions ask you to set up the problem but not to solve it, do not solve it. If the instructions ask you to use an equation but not to derive it, do not derive it.

A problem statement that asks: "Get the result in terms of y and z" asks you to solve for an equation with the result alone on the left hand side and nothing but numbers, constants, and y and z, or functions of y and z (y^3, $\sin z$, etc.) on the right hand side. If other quantities appear on the right hand side of the equation for the result, find their dependence on y and z and substitute into the result equation.

▽ ▽ ▽ ▽ ▽ ▽ ▽ ▽ ▽ ▽ ▽ ▽ ▽

Example 3: Oscillating car

When an automobile body is lowered onto its springs, the weight of the body will compress the springs a distance x. In a car with bad or no shock absorbers, when the body is pushed down on its springs it will bounce up and down (oscillate) at its resonant frequency of f cycles per second. Find f in terms of x and the gravitational constant g.

(The spring law is $F=-kx$ where F is the force that a spring exerts when it is compressed (or extended) a distance x, and k is the spring constant. The resonant oscillation frequency f of a mass m on a spring with spring constant k is

$$f = \frac{1}{2\pi}\sqrt{\frac{k}{m}} \quad .)$$

Solution for Example 3

Oscillating car

Name

Date

Drawing

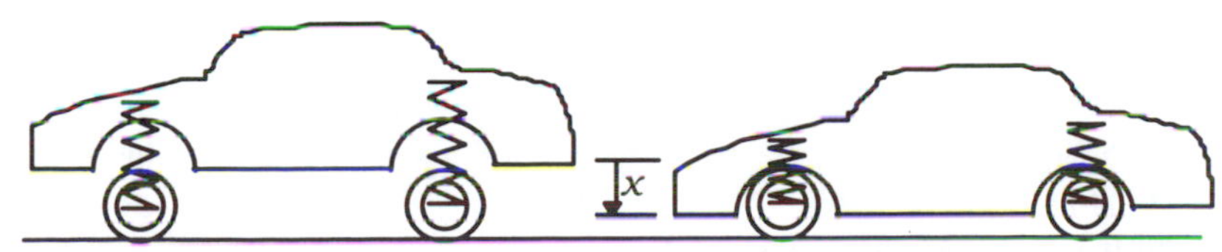

Definitions

Distance weight of body compresses springs	$= x$	m
Force exerted by springs to support body	$= F$	N
Weight of body	$= W$	N

Total force on body	$= F_T$	$= 0$	N
mass of body	$= m$		kg
gravitational constant	$= g$	$= 9.81$	m/s^2
spring constant for all of the car's springs acting together	$= k$		N/m
resonant frequency of body on springs	$= f$		cycles/ second

We know the gravitational constant g and the distance x the weight of the body compresses the springs. We do not know the spring constant k or the mass of the body m.

Try to find an equation that connects g, x, k, and m by writing the equations for the total force on the body when it is at rest and supported by its springs. The total force on the body is zero (it is not accelerating)

$$F_T = 0 \ .$$

The total force is the sum of the upward force exerted by the springs and the downward force exerted by the weight of the body. (We have taken the positive x direction as downward, so that the weight of the body is positive and the upward force exerted by the springs is negative.)

$$F_T = F + W \ .$$

From the previous two equations

$$0 = F + W$$

$$F = -W \ .$$

From the spring law

$$F = -kx \ .$$

Using the previous two equations

$$W = kx \ .$$

Connection between the weight of the body and its mass

$$W = mg \ .$$

Using the previous equations

$$kx = mg \ .$$

This equation contains the things we do know, g and x, and also the things we do not know, k and m.

Now let the car bounce up and down.

The resonant frequency is

$$f = \frac{1}{2\pi}\sqrt{\frac{k}{m}}$$

From the previous equation

$$\frac{k}{m} = \frac{g}{x}$$

so

$$f = \frac{1}{2\pi}\sqrt{\frac{g}{x}}$$

$$f = \frac{1}{2\pi}\sqrt{\frac{g}{x}}$$

The resonant frequency of the car body on its springs is 1 over 2π times the square root of the ratio of the gravitational constant to the distance the weight of the body compressed the springs.

△ △ △ △ △ △ △ △ △ △ △ △ △

Exercises

3–1. A 747 carrying 375 passengers flies at 950 km/hr and burns 10 metric tons of fuel each hour. If jet fuel costs $0.30 per kg, how much does it cost, per passenger, for the fuel necessary to fly the 4800 km from New York to Los Angeles?

3–2. Suzy uses floozles to cluse loozles. If it takes 25 floozles to cluse a dozen loozles, how many packages of 12 floozles will Suzy have to buy to have enough floozles to clooze 32 loozles? (Adapted from Grigory Oster's *Problem Book.*)

a. Define symbols for each quantity in this problem.
b. Don't forget to draw a really good diagram.
c. How many packages *will* she have to buy?

3–3. A copper wire carrying 120 kilowatts at 60,000 volts has a resistance that is its resistivity times its length divided by its cross-sectional area (its area perpendicular to its length.) Its mass is its density times its volume.

a. Draw a labeled diagram of the wire.
b. Define all the symbols in the problem.
c. Write an equation for the weight of the wire as a function of its length, radius, and density.
d. Write an equation for the resistance of the wire as a function of its length, radius, and resistivity.
e. Find the weight of the wire as a function of its resistance, length, resistivity, and density.

4

Describing the Problem

Describing a problem is the first step toward solving it. Describe a problem to understand it. Describe a problem to tell others what problem you have solved.

You cannot solve a problem you do not understand and you cannot understand a problem you have not described. Describe a problem by putting everything about the problem on paper in the form of drawings, symbol definitions, and data equations.

The next two sections are an example of the description of a problem and its solution. Parts of this example will be used in this chapter and the next.

▽ ▽ ▽ ▽ ▽ ▽ ▽ ▽ ▽ ▽ ▽ ▽ ▽

Example 4: Ultralight plane takeoff

An ultralight human-powered airplane and its pilot weigh 200 pounds. The pilot's pedaling drives a propeller that produces a forward force on the plane. The plane's motion through the air produces a force on its wings equal to a constant times the square of the plane's velocity. The direction of this air-flow force is 10° behind straight up, and the constant is 25 $N/(m/s)^2$. On takeoff, the pilot accelerates the plane along the runway with a constant acceleration of 0.6 m/s^2. The plane lifts off the ground when the upward component of the air-flow force on its wings equals its weight. How long a runway will the plane need to just lift off the ground?

Solution for Example 4

Ultralight plane takeoff

D. Scarl

11 August 1994

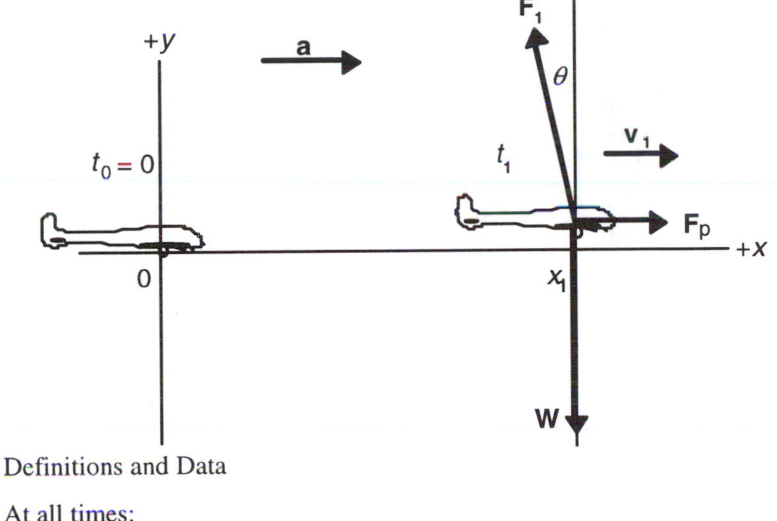

Definitions and Data

At all times:

weight of plane and pilot	$= W$	$= 200$	lb
acceleration along runway	$= a_x$	$= 0.6$	m/s^2
upward acceleration	$= a_y$	$= 0$	
air-flow force constant	$= K$	$= 25$	$\frac{N}{(m/s)^2}$
angle of force from vertical	$= \theta$	$= 10$	°

At beginning of runway

time at beginning	$= t_0$	$= 0$	s
distance along runway	$= x_0$	$= 0$	m
velocity	$= v_0$	$= 0$	m/s

At takeoff

time at takeoff	$= t_1$	s
distance along runway	$= x_1$	m
velocity	$= v_1$	m/s
force on plane from air flow	$= F_1$	N
upward air-flow force	$= F_y$	N
backward air-flow force	$= F_x$	N
force on plane from propeller	$= F_p$	N

Unit Conversion

$W \quad = 200 \text{ lb} \quad = 200 \text{ lb}\left(4.448\,\frac{\text{Newtons}}{\text{lb}}\right) \quad = 889.6 \text{ Newtons}$

Geometry:

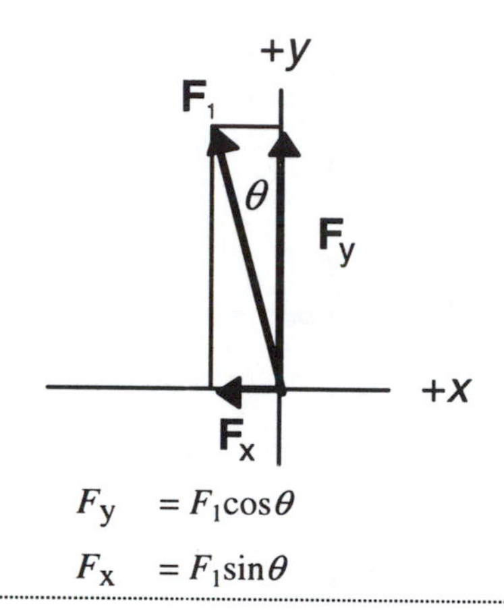

$F_y \quad = F_1\cos\theta$

$F_x \quad = F_1\sin\theta$

General equations

equations for constant acceleration

$\boldsymbol{a} \quad = \text{constant}$

$\boldsymbol{v} \quad = \boldsymbol{v}_0 + \boldsymbol{a}t$

$\boldsymbol{x} \quad = \boldsymbol{x}_0 + \boldsymbol{v}_0 t + \frac{1}{2}\boldsymbol{a}t^2$

connection between airflow force and speed

$F_1 \quad = Kv_1{}^2$

condition when plane just leaves ground

$F_y \quad = W$

Particular equations and vector components

Find v_1, the velocity at takeoff.

$F_1 \quad = Kv_1{}^2$

$v_1{}^2 \quad = \frac{F_1}{K}$

From the geometry equations

$F_1 \quad = \frac{F_y}{\cos\theta}$

$v_1{}^2 \quad = \frac{F_1}{K}$

$$= \frac{F_y}{K \cos\theta}$$

$$v_1 = \sqrt{\frac{F_y}{K \cos\theta}}$$

At takeoff the upward component of the airflow force equals the weight

$$F_y = W$$

$$v_1 = \sqrt{\frac{F_y}{K \cos\theta}}$$

$$v_1 = \sqrt{\frac{W}{K \cos\theta}}$$

$$= \sqrt{\frac{889.6 \text{ N}}{(25 \text{ N/m}^2\text{/s}^2)(\cos 10°)}}$$

$$= \sqrt{\frac{889.6 \text{ N}}{(25 \text{ N/m}^2\text{/s}^2)(0.985)}}$$

$$= \sqrt{36.13 \text{ m}^2\text{/s}^2}$$

$$v_1 = 6.01 \text{ m/s}$$

Find t_1, the time to reach takeoff speed.

For motion under constant acceleration

$$v = v_0 + at$$

$$v_1 = 0 + a_x t_1$$

$$t_1 = \frac{v_1}{a_x}$$

$$= \frac{6.01 \text{m/s}}{0.6 \text{m/s}^2}$$

$$t_1 = 10.02 \text{ sec}$$

Find x_1, the position at time t_1.

For motion under constant acceleration

$$x = x_0 + v_0 t + \frac{1}{2} a t^2$$

$$x_1 = 0 + 0 + \frac{1}{2} a_x {t_1}^2$$

$$= \frac{1}{2} a_x t_1^2$$

$$x_1 = \frac{1}{2}(0.6 \text{ m/s}^2)(10.02 \text{ s})^2 \qquad = 30.1 \text{ m}$$

Length of runway needed for takeoff $= x_1 = 30.1$ meters.

Check by getting algebraic solution

$$x_1 = \frac{1}{2} a_x t_1^2$$

$$= \frac{1}{2} a_x \left(\frac{v_1}{a_x}\right)^2$$

$$= \frac{1}{2} \frac{1}{a_x} (v_1^2)$$

$$= \frac{1}{2} \frac{1}{a_x} \left(\frac{W}{K\cos\theta}\right)$$

$$= \frac{889.6 \text{ N}}{2\ (25\text{N-s}^2/\text{m}^2)(0.6\text{m/s}^2)(\cos 10°)}$$

$$= \frac{889.6 \text{ N}}{2\ (25\text{N-s}^2/\text{m}^2)(0.6\text{m/s}^2)(0.985)}$$

$$x_1 = 30.1 \text{ m}$$

Length of runway needed for takeoff $= x_1 = 30.1$ meters.

△ △ △ △ △ △ △ △ △ △ △ △ △

Define the problem

The first part of the attack on a problem is to get all of the information about the problem on paper. This holds for homework problems, test problems, and especially for professional problems in which the problem itself usually is not clear at first.

Put a complete list of the facts of the problem on paper so that you can close the book or not look at the test paper again for that problem. Describe the problem so that someone who has never seen it can understand it. Do not copy the problem itself from the book or assignment sheet. Instead, write the facts as diagrams, definitions, data, and equations.

Even though the problem is stated clearly on your homework assignment or test, describe it again on paper before you try to solve it. Describing the problem on paper forces you to understand each part, it gathers everything together in one place, and it allows someone else to understand exactly what the problem is. Describing it on paper lets you start to work on the problem in an easy way without the terror of worrying about the answer or what equations to use. It is a good example of working instead of thinking.

Sometimes problems are not stated clearly in homework assignments, on tests, or in professional science and engineering. By describing the problem on paper before you begin to solve it you are training yourself to be able to do real problems.

Your description of the problem shows the person grading a test that you understood the statement of the problem. By writing the facts of the problem as you understand them, you are telling the person reading the solution what assumptions you made for those parts of the problem that were ambiguous. If you write your interpretation of the problem and make your assumptions clear, you can get credit for solving the problem using those assumptions.

Once the complete problem is on paper, go back, read the problem again, and check that you have every fact down correctly. Are those units centimeters or meters? If the motion started from rest, do you have $v_0=0$ on your paper? A missing fact at this point will make the solution harder to find and a wrong fact will make the solution wrong.

If you have a question about the statement of a problem on a test or on a homework assignment, describe the problem on paper, bring it to your instructor, and ask your question.

Write a heading

Title

Write a one-line title describing the problem.

▽ ▽ ▽ ▽ ▽ ▽ ▽ ▽ ▽ ▽ ▽ ▽ ▽

Ultralight Plane Takeoff
:
:

△ △ △ △ △ △ △ △ △ △ △ △ △

This first easy step in thinking about the problem lets you remember what the problem was about when you look at it a month later. Write the problem number as given in a book or assignment sheet. Then write the title, which tells you what the problem is about. Professional problems have no numbers, but they do have titles.

Name and date

Write your name and the date on every piece of written work that you do. Write your name, so that when you copy the work and show it to someone else they will know who the author was. Write the date, so that after you have solved a problem several times with increasingly good results, you will know which is the latest version.

Draw a diagram

Start with a diagram.

▽ ▽ ▽ ▽ ▽ ▽ ▽ ▽ ▽ ▽ ▽ ▽ ▽

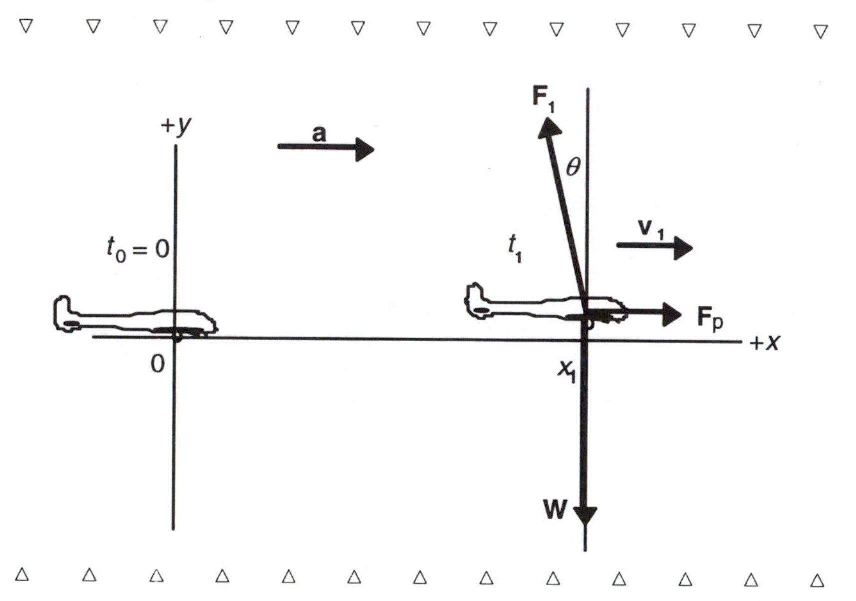

△ △ △ △ △ △ △ △ △ △ △ △ △

Most people think in pictures and find a diagram a help even in problems that have no obvious geometrical part. Venn diagrams in logic, organizational charts in management, and even outlines of written papers are diagrams that help you to think. Using diagrams to describe scientific and engineering problems will help you develop the ability to translate words into pictures and pictures into words.

Diagram or picture

A picture can be artistic, subjective, and abstract, suggesting objects or emotions. A diagram, on the other hand, shows, realistically and to scale, the objects that are part of a problem. It also shows abstract symbols such as axes and vectors and the symbols and numbers that describe the properties of objects, such as position, mass, velocity, and acceleration.

If you draw a realistic and accurate diagram, you will find it easier to understand the problem. In a problem about a car, a diagram that looks at least a little like a car is easier to understand than a simple rectangle.

Draw your diagrams large so that all the objects, symbols, and numbers are clear. For a simple diagram 50 mm high and 100 mm wide is a good start. Complicated drawings may deserve a whole page. In this book and others diagrams are small to save space. Hand-drawn diagrams need to be much larger than those in books.

Draw diagrams to scale to make it easier to understand what is going on. For instance, if at t_1 a car is 2 meters from the origin and at t_2 it is at 20 meters, make sure that x_2 is about 10 times bigger than x_1. You can make this easier by drawing an x axis divided into intervals, as described in the next section.

Some problems need several diagrams. Use one to show the initial conditions, another to show the final conditions, one on which to draw the trigonometric relationships, and perhaps one to display the answer. In problems that have things happening at different times, draw a diagram at each of the times the problem talks about, like a multiple-exposure snapshot.

When you do not know enough about the answer to draw a correct diagram, just make reasonable assumptions and draw a diagram anyway. If the diagram is not right, the answer will tell you so, by giving a negative number, for instance.

Try to draw a diagram that is not a special case. If the problem is about a triangle, draw a general triangle, not one that includes a right angle or one that has equal sides. A diagram of a pendulum at a general place in its swing is more useful than one with the pendulum at the top or bottom of its swing.

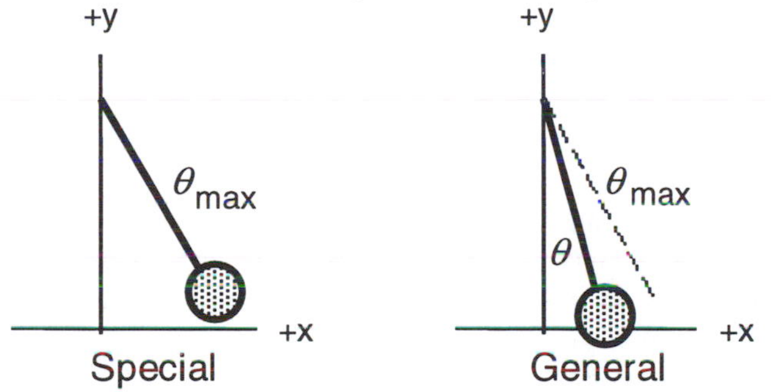

On the diagram on the right both the angle θ of the pendulum and the special angle θ_{max} can be shown. On the diagram on the left, θ cannot be shown.

One trick that helps with complicated problems is to draw a first diagram showing the problem with all of its known quantities and a second diagram showing the solution with its unknown quantities. This works well with electromagnetic field problems when the first diagram shows the electrodes or coils that cause the field and the second diagram shows the field caused by those electrodes or coils.

Axes

Most diagrams need a set of axes. Draw x and y axes and label the $+x$ and $+y$ directions and the zero position.

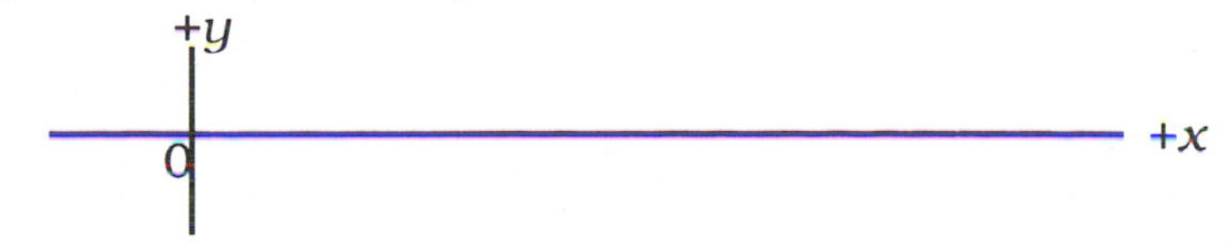

If the description of the problem includes positions along the axis, label each axis with axis divisions before you start to draw the description of the problem. For instance, if the problem involves a copper ring whose diameter is 2 mm and whose center lies on the +x axis 5 mm from the origin, the axes would look like this.

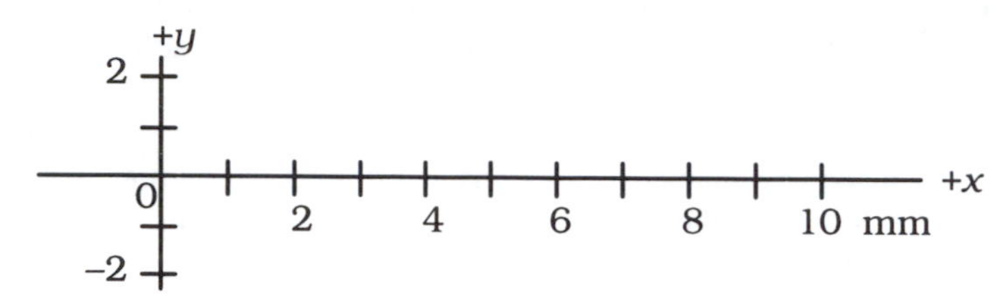

The ring would look like this.

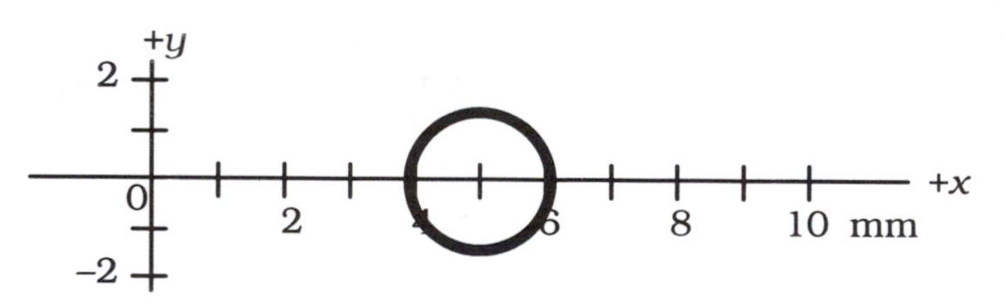

By drawing and labeling the axes first it is much easier to draw the ring accurately. Of course, before you drew the divisions on your axes, you used the known size and position of the ring to decide on the size of the divisions.

When putting a scale along an axis, divide the scale in steps of 1, 2, or 5 times a power of 10. For instance, good scales are 0.3, 0.4, 0.5, 0.6 ..., 22, 23, 24, 25 ..., 0, 50, 100, 150 ..., and so forth. This makes it easy to find values that fall between the divisions.

Once you have an accurate scale drawing, finding any distance, for instance, the distance from the origin to the top of the ring, is easy.

Scale

Notice that although the diameter of the ring is only 2 mm, the drawing is much larger. The way chemists get drawings of molecules and architects get drawings of buildings to fit on a standard sheet of paper is to draw them to scale. The drawing of the ring is at a scale of 10:1. Ten millimeters on the paper stands for 1 mm in the real world. Label the axes on the drawing with real-world distances, not lengths on the paper.

Three-dimensional drawings

Drawings on paper are two-dimensional but real-world objects are three-dimensional. When an object, like the ring above, is simple and when all the vectors in the problem lie in one plane, it is easy to describe the problem with a single drawing. However, when the object is complicated or the interesting dimensions or vectors do not lie in a plane, it takes more than one view to describe the problem clearly.

A standard way to draw an object is to draw three views, one from the side, one from the top, and one from the front. The top view, for instance, is what you would see if you looked down on the object from above. Axis lines help to align the three drawings. Draw the axes first, then the side or front view, then the top view directly above, and then the front or side view to the right.

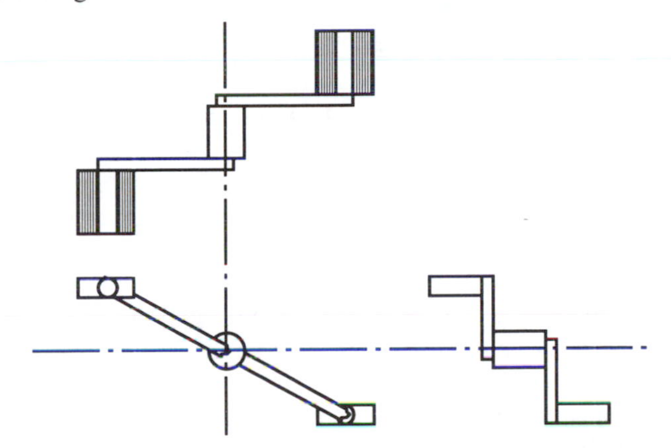

In these three views of a bicycle crank, the position and width of the pedals in the side view sets the position and width in the top view. You can use light lines extending straight up from the side view to make sure that the position and width of the pedals are the same in the top view as in the side view. In the same way you can use light lines extending toward the right to help make the position and height of the pedals the same in the front view as in the side view.

Labels

Since the meaning of a symbol is clearer when it appears on a diagram, include all the appropriate symbols on your diagram. When you know the values of quantities like times, distances, masses, and forces, the problem will be easier to understand if you show both symbols and numbers. For instance, the diameter of the ring above could be shown as d = 2 mm.

The labeled diagram should show at a glance what the problem is about. If you draw the diagram to scale and put the known numbers on the diagram, the problem will become more real and easier to think about.

The symbols and numbers on the diagram will appear again in the table of definitions and data.

Drawing vectors

Any vector component pointing in the $+x$ or $+y$ direction on a set of axes is positive. A vector component pointing in the $-x$ or $-y$ direction is negative.

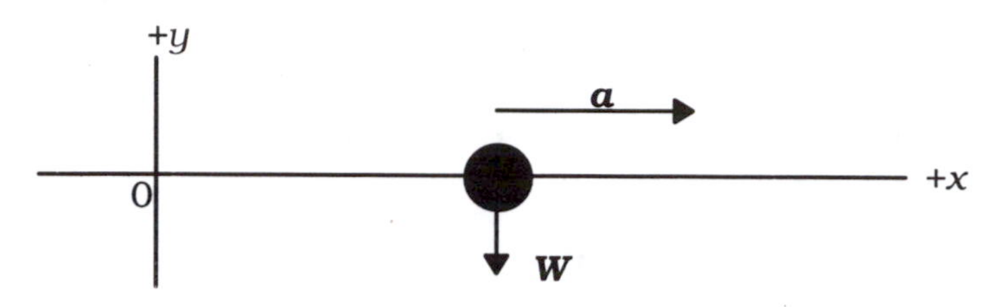

In this drawing, although the vectors themselves have no sign, the x component of **a** is positive and the y component of **W** is negative.

Although the x and y axes have to be perpendicular, they do not have to be horizontal and vertical. When the problem involves known vectors, the wisest choice of axis directions is the one that aligns the axes with as many of the given vectors as possible. For instance, in a simple inclined plane problem, if you choose one axis along the plane and the other axis perpendicular to the plane, all of the forces except the weight will point along one of the axes.

Although forces can push or pull, it is a good convention to draw all force vectors as if they pull. Draw an X on the object at the point at which the force acts and then draw the force vector with its tail starting at the X. This convention simplifies calculations of distances times forces to find quantities like work and torque.

To avoid confusing velocity and acceleration vectors with force vectors, draw velocity and acceleration vectors near the object they describe but not touching it.

Name the variables

Each number that enters into the problem needs to be given a symbol. Each constant, each variable, each initial condition, and each result needs a symbol. Until you define a symbol for a quantity you cannot talk about that quantity, define it exactly, use it in an equation, or even think about it efficiently.

Although a quantity is given as a *number* in the statement of the problem, give it a *symbol* when you work on the problem. Use the number only in the definition of the quantity and then again in the last few steps of the solution when you put in numbers to find the answer.

In most problems the name of each quantity is a symbol—a letter decorated with subscripts, like v_0. In a spreadsheet or computer program the name can be several letters that describe the quantity, *INITVEL*.

Sometimes it is hard to notice that a number given in the statement of a problem needs a symbol. For instance, if the mass of object B is twice the mass of object A, the best way to describe that fact might be

ratio of mass B to mass A	= s	= 2

and, in the preliminary equations

$$\frac{m_B}{m_A} = s.$$

Or, if the problem states, *The mass of B is 10% greater than the mass of A*, the symbol might be defined as

ratio of mass B to mass A	= s	= 110%	= 1.1

or

ratio of (m_B–m_A) to m_A	= u	= 10%	= 0.1 .

If you do not define a symbol for these ratios and later need to calculate the result for a new ratio, you will have to redo the whole problem. If you use the symbol and put in the number only at the end, you will be able to find the new result simply by putting in the new number.

Symbol definitions

Make a list of *all* of the quantities that enter the problem.

▽ ▽ ▽ ▽ ▽ ▽ ▽ ▽ ▽ ▽ ▽ ▽ ▽

Definitions and Data

At all times:

weight of plane and pilot	= W	= 200	lb
acceleration along runway	= a_x	= 0.6	m/s^2
upward acceleration	= a_y	= 0	
air-flow force constant	= K	= 25	$\frac{N}{(m/s)^2}$
angle of force from vertical	= θ	= 10	°

At beginning of runway:

time at beginning	= t_0	=0	s
distance along runway	= x_0	=0	m
velocity	= v_0	=0	m/s

At takeoff:

time at takeoff	= t_1	s
distance along runway	= x_1	m
velocity	= v_1	m/s
force on plane from air flow	= F_1	N
upward air-flow force	= F_y	N
backward air-flow force	= F_x	N

force on plane from propeller $= F_p$ N

△ △ △ △ △ △ △ △ △ △ △ △ △

Define every given quantity, every initial condition, every useful intermediate quantity that you can use to help solve the problem, and every unknown quantity that will be one of the results of the problem.

You can organize your definitions in any way you want. One useful way is to define first those quantities that remain constant, then those quantities that describe what is happening at t_0, then those that describe what is happening at t_1, and so on. At each time you may need to define angles, positions, velocities, accelerations, forces, momenta, energies, and other quantities. Start with the simplest of these quantities and build up to the more complicated.

Another way to organize definitions is first to define the quantities whose values you know and then the quantities whose values you do not know. This order makes the structure of the problem clear but means that some similar properties, such as the horizontal and vertical initial positions of an object, may be separated in the list of definitions.

Write the unit for each quantity. It is amazing that just writing the unit for a mysterious quantity makes it less mysterious and more understandable. It also avoids mistakes when you notice that the quantity is given in centimeters and you are working in meters.

Remember to label the diagram with the symbols that you have made up for each quantity.

▽ ▽ ▽ ▽ ▽ ▽ ▽ ▽ ▽ ▽ ▽ ▽ ▽

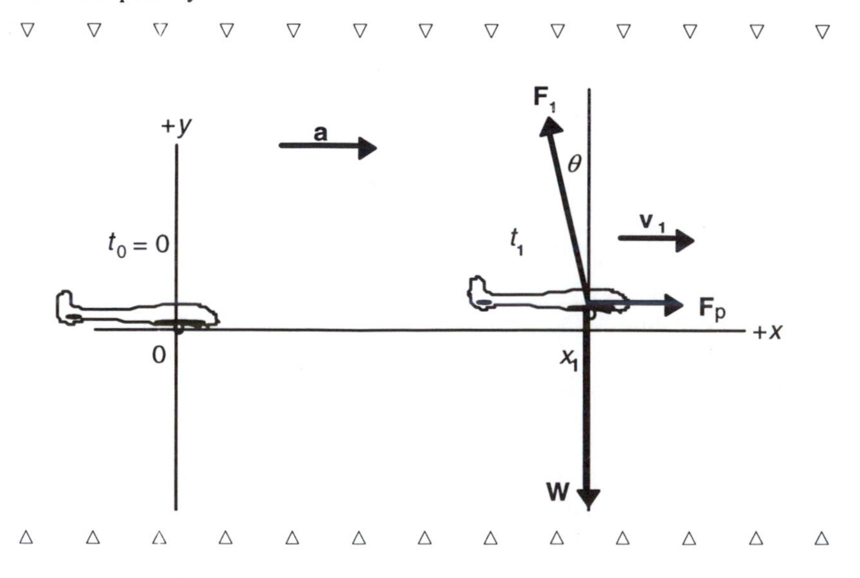

△ △ △ △ △ △ △ △ △ △ △ △ △

Zero values

Quantities that are zero are easy to forget or disregard but they have all the rights and privileges of other quantities. If an object is moving with constant velocity, its acceleration is zero and that fact is an important part of the statement of the problem. You need to give that acceleration its own symbol, definition, and value. Write, at the beginning of the data equations

acceleration = a = 0 m/s^2 .

Whenever a problem gives a time, the position of an object, or the direction of a line, that time, position, or direction is measured with respect to a zero time, position, or direction. You need to define where the zero is before the time, position, or direction itself has any meaning.

Zero values are often described in a way that makes them hard to notice. *Initially*, *at the origin*, *at rest*, and *dropped* are all disguises for quantities that are zero. Find that quantity and give it a symbol and definition.

You can draw zero-length vectors on the diagram describing the problem. Just draw a short vector and label it to show that the value of the quantity it represents is zero.

Choosing symbols

If the symbols for the quantities you need are given in the problem or are famous, choosing symbols is easy. Use the same symbols your text and your teacher use for time, mass, position, velocity, acceleration, force, energy, momentum, torque, and so on.

If the symbols for the quantities you need are not given, make them up. Here are some generally accepted rules for choosing symbols

x,y,z	distances
i,j,k,l,m,n	integers
a,b,c	constants
$\alpha,\beta,\gamma,\theta$	angles .

Upper case letters are often constants, while lower case letters are often variables (except for a, b, and c). For instance, the radius of the earth is often R, while the distance from the center of the earth to any point inside or outside the earth is often r.

Subscripts

Symbols like v for velocity need to be decorated with subscripts when there are several different velocities in the problem. Subscripts describe *which* velocity you are talking about.

Subscripts that are numbers usually indicate that the quantity is a constant.

initial velocity = v_0 = 0 m/s

(Although v may change with time, v_0 is a constant.)

The best way to label times is with numerical subscripts, so that all the variables at a given time have the same subscript. x_1, v_1, and p_1 are the position, velocity, and momentum at time t_1; they are constants, and each has only a single value. x, v, and p are the position, velocity, and momentum at any time t; they are variables and have different values for every value of t.

Subscripts that are lower case letters usually indicate that the quantity is a variable.

velocity in the x direction = v_x

(v_x changes with time. Notice that v_{x0} or v_{0x} is the initial velocity in the x direction and is a constant.)

Subscripts that are capital letters usually describe different objects. M_A and M_B are the masses of object A and object B.

Although these rules will be useful when you start choosing symbols, you will eventually need to break almost every one of them. Any symbol that you define and use consistently will work fine.

Data equations

At the beginning of a problem sometimes all you have is a quantity's definition. However, if you have a *number*, include it in the definition. When you include a number in the definition, the definition becomes a data equation.

▽ ▽ ▽ ▽ ▽ ▽ ▽ ▽ ▽ ▽ ▽ ▽ ▽

Definitions and Data

weight of plane and pilot	$= W$	$= 200$	lb
acceleration along runway	$= a_x$	$= 0.6$	m/s^2
upward acceleration	$= a_y$	$= 0$	
air-flow force constant	$= K$	$= 25$	$\frac{\text{N}}{(\text{m/s})^2}$
angle of force from vertical	$= \theta$	$= 10$	°

△ △ △ △ △ △ △ △ △ △ △ △ △

The data equations turn the written statement of the problem into a set of equations. Read each phrase of the problem and write down the equation it generates. The sentence, *A ball is dropped from a height of 7 meters,* generates these two data equations

initial height	$= y_0$	$= 7$	m
initial velocity	$= v_0$	$= 0$	m .

The second equation comes from the simple word *dropped.*

Watch for these phrases and the equations they generate

starts from rest

initial velocity	$= v_0$	$= 0$	m/s

moves with constant velocity

acceleration	$= a$	$= 0$	m/s^2

two equal masses

mass of body 1	$= m_1$		kg
mass of body 2	$= m_2$	$= m_1$	kg

Sometimes a definition and a data equation look almost alike.

Definition

$$F = F_1 + F_2$$

Data Equation

$$F = 0.$$

The definition equation $F=F_1+F_2$ defines the total force F as the sum of force F_1 and force F_2. The data equation says that for some reason the total force F is zero. It is tempting to combine these two different equations into one equation

$$F_1 + F_2 = 0.$$

But now you have no symbol for the total force, and it may not be at all obvious why the sum of F_1 and F_2 is zero. Be careful to separate definition equations and data equations and to write down both .

In your data equations and calculations every equation has to have a left hand side. Writing down *ma* by itself doesn't mean anything at all.

Check

After drawing a diagram and writing down all the symbol definitions and data equations that the statement of the problem generates, read the problem again. Check that every phrase has its own equation. Check your definition of each quantity against the definition in the problem statement. Check your unit for each quantity against the unit in the problem statement.

Check that you have chosen a symbol for the quantity that will be the answer, even though you have no idea what the answer is.

Write preliminary equations

After the data equations you will need the preliminary equations that prepare the data for the science equations. The preliminary equations change units, record the values of constants you have looked up, calculate components of vectors, and so forth.

Since it is better not to complicate the original diagrams, preliminary calculations sometimes need their own new diagrams. The original diagrams defined the problem. The new ones are part of your preparation for solving the problem.

Units

There is no longer any choice of units. All of your courses will use the meters, kilograms, seconds (SI) system. If you live in the United States and become a civil or automotive engineer, an architect, a machinist, or a carpenter, you can learn to use American units on the job.

If a quantity is given in non-SI units, change to SI. Change units in a way that can be checked later. Write the quantity with its original unit, an equals sign, the original quantity again, the multiplication factor that changes the unit, an equals sign, and the quantity with the new unit.

▽ ▽ ▽ ▽ ▽ ▽ ▽ ▽ ▽ ▽ ▽ ▽ ▽

$$\begin{aligned} W &= 200 \text{ pounds} \\ &= 200 \text{ pounds} \left(\frac{4.448 \text{ newtons}}{1 \text{ pound}} \right) \\ &= 889.6 \text{ newtons} \end{aligned}$$

△ △ △ △ △ △ △ △ △ △ △ △ △

There are, unfortunately, some exceptions to this rule. If the problem is given entirely in non-SI units, and the answer is acceptable in non-SI units, solve the problem with the given units. If the problem is stated in SI units but you are asked for the answer in non-SI units, calculate the solution in SI units and then convert the answer at the very end. These cases still arise on tests, but less and less often in scientific and engineering practice.

Constants

Write down the constants that you will need to solve the problem. Sometimes the value of these constants will be given in the statement of the problem and sometimes you will have to look them up. For each constant whose value you look up write the name of the book or paper in which you found the value. If the constant is a rare one and hard to find, you will appreciate the reference when you come back to it later.

▽ ▽ ▽ ▽ ▽ ▽ ▽ ▽ ▽ ▽ ▽ ▽ ▽

earth's gravitational constant $= g$

$$\begin{aligned} g &= 980.665 \text{ cm/s}^2 \\ &= 980.665 \text{ cm/s}^2 \left(\frac{1 \text{ m}}{100 \text{ cm}} \right) \\ &= 9.807 \text{ m/s}^2 \end{aligned}$$

From C. W. Allen *Astrophysical Quantities,* pg. 109, Athlone Press, London 1955.

△ △ △ △ △ △ △ △ △ △ △ △ △

Trigonometry and vector components

Geometry complicates science and engineering problems. Doing the geometrical calculations first simplifies the solution by separating the geometry from the science.

▽ ▽ ▽ ▽ ▽ ▽ ▽ ▽ ▽ ▽ ▽ ▽ ▽

Geometry

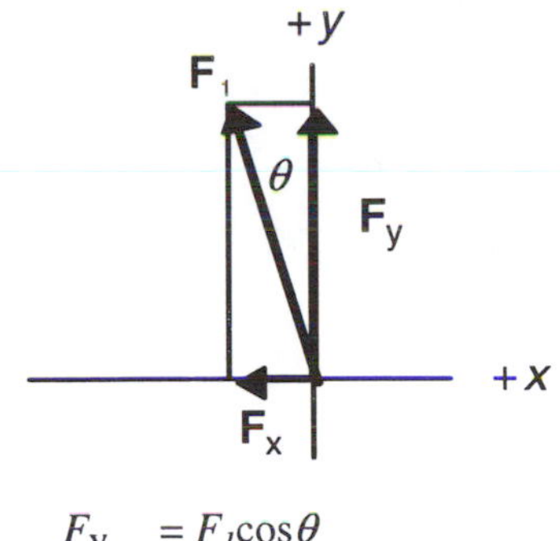

$$F_y = F_1 \cos\theta$$

$$F_x = F_1 \sin\theta$$

△ △ △ △ △ △ △ △ △ △ △ △ △

Especially in three-dimensional problems, defining and describing positions, angles, and directions are sometimes the most difficult part of the problem.

Exercises

4–1. Two iceskaters are standing in the middle of a 20 m by 30 m rink whose long axis runs east–west. One skates east and the other skates west; at 1 second they are each 4 m from their original position. At 2 s they are each 9 m from their original position and are each moving with a speed of 5 m/s.

a. Draw a labeled diagram of the skaters and rink.

b. Define all of the quantities given in the problem.

4–2. A car heading east on Route 66 at 55 mph starts to brake for a red light. At t=0 the car's acceleration is 3 m/s^2 toward the west.

a. Draw a labeled diagram, with the positive x axis toward the east.

b. Draw the car's velocity and acceleration vectors.

c. Define all of the quantities mentioned in the problem.

d. What is the sign of the x component of the car's velocity? What is the sign of the x component of the car's acceleration?

4–3. A hot–air balloon 35 m above a barking dog on the ground is moving toward the north–east while it is ascending at 3 m/s. The balloon's total velocity is 10 m/s.

a. Draw at least two labeled diagrams showing the balloon's position and velocity.

b. Define all of the quantities given in the problem.

4–4. The nearest star, Alpha Centauri, is 4.3 light–years away.

a. How fast, in meters/second, must a spaceship from earth travel in order to reach Alpha Centauri in a human lifetime? (Neglect relativistic effects.)

b. What is the ratio of the velocity of the ship to the velocity of light?

5

Finding the Solution

Science equations

You have stated the problem exactly, labeled the drawings, written the data equations, converted units, made auxiliary drawings, found the needed constants, and done the trigonometry.

Now you are ready to think about solving the problem. Notice that until now you have not had to think. All of the preparation you have done is the same for every problem and will soon become automatic.

Rather than beginning at the beginning, amateur problem solvers often begin at this point and then wonder why problem solving is so difficult and why their exam scores are so low. Instead, you have been doing all of the standard preparation, while thinking about the problem in the back of your mind. You have all of the information about the problem on paper. On an exam you have impressed the test grader that you know something about the problem, and you have already gotten about one-quarter credit for it, all without knowing how to do the problem.

Choosing the right equation

What kind of thinking do you need? Not the kind that tries to solve the problem in one bite, but the thinking that asks, "What is going on here? Is the velocity constant? Is the acceleration constant? Are the forces constant? Is momentum conserved? Is energy conserved? Is angular momentum conserved? Is the body in free fall? Is it going in a circle? At what time in its motion do I have information I can use? How many different times are involved?" The information you already have on paper will provide new knowledge that will let you solve the problem.

Classify the problem

When you classified and described each equation as you learned it, you prepared yourself for choosing the equations that will let you solve the problem.

You have classified equations; now classify the problem. Does the problem involve

motion with constant acceleration
motion in a circle
forces in equilibrium
forces that change with time
motion in two dimensions
motion under a gravitational force

or one of many other categories?

Compare your classification of the problem with your list of classified equations to eliminate all equations but a few. Of these choose the equations that connect quantities given in the data equations with each other, equations that connect the desired results with each other, or equations that connect the data with the results.

If you know which chapter of your textbook a homework or exam problem comes from, it is likely that the equations you need will come from the same chapter. Of course, that way of choosing equations will not be too useful in later professional problems.

Classify quantities

When you write down an equation that is a candidate for your problem, underline the quantities in the equation that are given in the statement of the problem or are known constants. Circle the quantity that is asked for as one of the results of the problem. Then count the number of quantities that are neither given nor asked for. If the number of extra quantities is zero, you have chosen the right equation, but check again that the equation holds under the conditions of the problem. If you have one extra quantity but the extra quantity appears in two different equations, you can eliminate it and solve the problem. If you have more extra quantities than equations, look for another equation that holds under the conditions of the problem or reread the problem to make sure that you did not miss a quantity whose value was given. (Did you write down $v_0 = 0$ when you read *starts from rest*?)

You do not have to start with a single equation that connects the data directly with the result. What you need is a chain of equations, each of which depends on the next. One equation should contain each useful piece of data and one should contain each unknown.

This is not the place to worry about whether you can do the algebra that will eventually connect the data with the result. That comes later.

General equations

Before you write any equations, write a few words saying what you are trying to find. This is part of the general rule of saying what you are going to do before you start doing it. These words help you and whoever else reads your work to understand the flow of what you are doing. Write *Calculate the distance at time* t_2, or *Find the time at which the velocity is zero* before you start to do those things.

Before you write down an equation, write the reason that you chose it. If the acceleration of an object in the problem is constant, and you are going to use one of the equations that applies to motion under constant acceleration, write *for constant acceleration*. If energy is conserved, write *since energy is conserved*.

Write down these conditions so that they can help you to think about the problem and so that you can check them later for errors. The assumption that you do not write down will be the one that is wrong.

Write each equation first in its original form, that is, the form in which it is famous or in the form in which you found it. Just by writing an equation in its famous form on a test you can get credit for knowing something.

In professional work write the reference to the book or paper in which you found each equation, since you may not remember where the equation came from when you look at your solution months later.

▽ ▽ ▽ ▽ ▽ ▽ ▽ ▽ ▽ ▽ ▽ ▽ ▽

For constant acceleration

$$\mathbf{a} = \text{constant}$$

$$\mathbf{v} = \mathbf{v}_0 + \mathbf{a}t$$

$$\mathbf{r} = \mathbf{r}_0 + \mathbf{v}_0 t + \frac{1}{2}\mathbf{a}t^2$$

△ △ △ △ △ △ △ △ △ △ △ △ △

In this case the general equations are vector equations. Unless you can solve the problem by using vector algebra, a general vector equation needs to be written as three particular equations for the three components of the vectors. The next section shows an example of this.

Particular equations

After you have written the conditions under which an equation holds and the general, famous form of the equation, write the equation again using the particular variables that you have defined for your problem. Now x becomes x_1 or x_2 or even d, y becomes y_1 or h, and so on.

▽ ▽ ▽ ▽ ▽ ▽ ▽ ▽ ▽ ▽ ▽ ▽ ▽

For constant acceleration

$$\mathbf{v} = \mathbf{v}_0 + \mathbf{a}t$$

$$\mathbf{v}_2 = \mathbf{v}_0 + \mathbf{a}t_2$$

△ △ △ △ △ △ △ △ △ △ △ △ △

For vector equations write separate equations for the components of the vectors.

▽ ▽ ▽ ▽ ▽ ▽ ▽ ▽ ▽ ▽ ▽ ▽ ▽

For constant acceleration

$$v_x = v_{0x} + a_x t$$

$$v_y = v_{0y} + a_y t$$

$$v_z = v_{0z} + a_z t$$

$$x = x_0 + v_{0x}t + \frac{1}{2} a_x t^2$$

$$y = y_0 + v_{0y}t + \frac{1}{2} a_y t^2$$

$$z = z_0 + v_{0z}t + \frac{1}{2} a_z t^2$$

△ △ △ △ △ △ △ △ △ △ △ △ △

Algebra

After you have written the equations that connect the known quantities with the unknown quantities in a problem, you could use a computer program like Maple® or Mathematica® to combine the equations and solve for the unknowns. However, most teachers still require you to do the solution yourself on homework and tests, so you have to be able to do algebra accurately and well.

To use algebra well you need to be able to do three things: know the common algebraic operations, figure out which operations to use to connect known and unknown quantities, and copy symbols correctly from one step to the next

Use a small number of known operations

It is easy and useful to make a short list of all of the algebraic operations that you will ever need: multiplying both sides of an equation by something, adding something to both sides of an equation, multiplying a sum of terms by another sum of terms, and so forth. Eliminate the operations that are made up of a series of simpler operations. (One step at a time.) The remaining operations are your algebraic tool kit. Use only these operations for the rest of your life.

Many troubles with algebra come from sudden on-the-spot invention of new algebraic operations. If, in the heat of solving a problem, you suddenly decide that

? $\frac{1}{a+b} = \frac{1}{a} + \frac{1}{b}$?

you have invented an operation that does not exist. Stick to a small number of algebraic operations, do them one at a time, and be sure not to invent new ones, no matter how useful they may seem.

Get unknowns on left hand side

The goal is to get each unknown by itself on the left hand side of an equation with the right hand side of the equation containing only quantities whose values are known.

This usually has to be done in several steps. First get the unknown on the left hand side by itself.

▽ ▽ ▽ ▽ ▽ ▽ ▽ ▽ ▽ ▽ ▽ ▽ ▽

$$x_1 = \frac{1}{2} a_x t_1^2$$

△ △ △ △ △ △ △ △ △ △ △ △ △

Then use other equations to get rid of intermediate unknowns on the right hand side.

▽ ▽ ▽ ▽ ▽ ▽ ▽ ▽ ▽ ▽ ▽ ▽ ▽

$$x_1 = \frac{1}{2} a_x t_1^2$$

$$= \frac{1}{2} a_x \left(\frac{v_1}{a_x}\right)^2$$

$$= \frac{1}{2} \frac{1}{a_x} (v_1^2)$$

$$= \frac{1}{2} \frac{1}{a_x} \left(\frac{W}{K\cos\theta}\right)$$

△ △ △ △ △ △ △ △ △ △ △ △ △

There are two kinds of unknown quantities. The first kind is one of the results you need.

▽ ▽ ▽ ▽ ▽ ▽ ▽ ▽ ▽ ▽ ▽ ▽ ▽

To find x.

For motion under constant acceleration

$$x = x_0 + v_0 t + \frac{1}{2} a t^2$$

$$x_1 = 0 + 0 + \frac{1}{2} a_x t_1^2$$

$$= \frac{1}{2} a_x t_1^2$$

△ △ △ △ △ △ △ △ △ △ △ △ △

The second kind of unknown is an intermediate result that you need to find before you can calculate one of the results you need.

▽ ▽ ▽ ▽ ▽ ▽ ▽ ▽ ▽ ▽ ▽ ▽ ▽

To find t

For motion under constant acceleration

$$v = v_0 + at$$

$$v_1 = 0 + a_x t_1$$

$$t_1 = \frac{v_1}{a_x}$$

$$= \frac{6.01 \text{ m/s}}{0.6 \text{ m/s}^2}$$

$$= 10.02 \text{ sec}$$

△ △ △ △ △ △ △ △ △ △ △ △ △

Each intermediate result can be handled two different ways. You can leave it as an algebraic expression ($t{=}v_1/a_x$ in the example above) or you can calculate its numerical value ($t{=}10.02$ s in the example above.) Each choice has some advantages.

You can get a completely algebraic final result by putting the algebraic expression for an intermediate result into the equation for a final result. If some quantities cancel each other, the equation for the result becomes simpler. Sometimes you can learn that the result does not even depend on one of the quantities given in the statement of the problem. Simplifying the result in this way helps in understanding the problem. In addition, if the numbers given in the description of the problem change, you will only have to recalculate the values of the final results at the end of your solution and not the values of the intermediate results that may be scattered all through your calculation.

On the other hand, it is often worthwhile to calculate intermediate results, find their values, and think about whether they are reasonable. This is one place where you can enter numbers slightly before you get to the end of the problem. If an intermediate result will be used several times in the rest of the solution, calculating its numerical value the first time it appears can avoid having to calculate it again each time it appears. Calculating intermediate quantities provides a useful check and also simplifies the equations for the results.

Since calculating the numerical values of intermediate results is useful, and keeping the algebraic expressions for intermediate results is also useful, experienced problem solvers often do both. If their final equations simplify, good. If not, they go back and use the numerical values.

Copy symbols exactly

Copy symbols correctly from one line to the next. Things to watch for are

Exponents	It is easy to write x instead of x^2. This is especially dangerous when the quantity is in the denominator. If x=5 and you write, in the last line of a problem $y = \frac{1}{x^2} = \frac{1}{5} = 0.2,$ it can spoil your whole problem and, maybe, your day.
Subscripts	It is easy to write x_a instead of x_b,
Negative signs	– is small and inconspicuous. Treat it with respect. Even the most experienced scientists and engineers have trouble here.

Checks

Check every calculation.

Experienced problem-solvers check each line of a calculation immediately after they write it. They check the algebra, they check the unit, they do a variational check, and they check whether the result is reasonable. If a calculation has ten steps and the probability of getting each step right is 95%, the probability of getting the whole calculation right is only 60%. Get those probabilities up by checking each step before going on to the next.

A good way to check a calculation is to do it again in an entirely different way. If you do not know of a different way to do it, check by redoing it the same way, but realize that it is very difficult to see mistakes once they are made. Everything usually looks all right.

Check algebra

Check each algebraic step. Do each one a different way if you can. Check that you have not done two steps at once. Be careful with exponents, subscripts, and negative signs.

Check units

A good check for algebraic mistakes is to look at the units of all of the quantities in the symbol equation for each result. If your original equations are right but the units on the right hand side of your result equation are not the same as the units on the left hand side, you have probably made a mistake in algebra. In fact, by dividing the units on the left hand of the equation by the units on the right, you can sometimes find the units of the missing or extra factor. This is one of the few checks that not only tells you that something is wrong, but tells you *what* is wrong.

Make a variational check

Check the algebraic form of each result while it still contains symbols before you put the numbers in. Check how the left hand side changes when quantities on the right hand side change.

▽ ▽ ▽ ▽ ▽ ▽ ▽ ▽ ▽ ▽ ▽ ▽ ▽

$$x_1 = \frac{1}{2} a_x t_1^2 = \frac{1}{2} a_x \left(\frac{v_1}{a_x}\right)^2$$

$$= \frac{1}{2} \frac{1}{a_x} (v_1^2)$$

$$= \frac{1}{2} \frac{1}{a_x} \left(\frac{W}{K\cos\theta}\right)$$

△ △ △ △ △ △ △ △ △ △ △ △ △

Should the runway length x_1 increase as the weight W increases?

Should the length increase as the angle θ between the force and the vertical increases?

Should the length decrease as the force constant K increases?

Should the length decrease as the acceleration down the runway a_x increases?

If the variation does not seem right, there may be an algebraic mistake.

Check test cases

Put in special test values of the numbers, not the ones in the actual problem, to see if the result makes sense with these test values.

Zero

Zero is one of the best test values. Think about what should happen to the result if one variable goes to zero. Then put zero in for that variable, and see if the result does what you expect. Repeat for the next variable.

Infinity

Sometimes infinity is a good test value. What should happen to the result if one of the variables goes to infinity? Should it get very large or very small? Put in infinity for that variable and see if the result does what you expect. Do the same for the other variables.

Check by approximation

Check your arithmetic by doing it approximately. Many experienced scientists and engineers have learned to do addition, subtraction, multiplication, division, and even square roots approximately. They often do this in their heads, but you are safer if you write down the approximate calculation just like any other calculation.

To do an approximate calculation, round all of the numbers to one easy digit times a power of ten. 1.4 becomes 1, 1.7 becomes 2, any number between 7 and 15 becomes 10, 24 becomes 20, 72 becomes 1×10^2, and so forth. This preserves an accuracy of about 50%. Since rounding up some numbers compensates for rounding down others, the accuracy of the answer will usually be better than 50%.

Now do the calculation using your rounded numbers without a calculator. Compare your approximate answer with your original accurate calculation. If the two differ by more than about 50%, check both the accurate and the approximate calculations. This is a good way to catch errors in exponents and mistakes in entering numbers in your calculator.

▽ ▽ ▽ ▽ ▽ ▽ ▽ ▽ ▽ ▽ ▽ ▽ ▽

$$x_1 = \frac{1}{2}\frac{1}{a_x}\left(\frac{W}{K\cos\theta}\right)$$

$$= \frac{889.6\ \text{N}}{2\,(25\ \text{N-s}^2/\text{m}^2)(0.6\ \text{m/s}^2)(\cos 10°)}$$

$$= \frac{889.6\ \text{N}}{2\,(25\ \text{N-s}^2/\text{m}^2)(0.6\ \text{m/s}^2)(0.985)}$$

$$x_1 = 30.1\ \text{m}$$

Check by approximate calculation.

$$x_1 = \frac{889.6\ \text{N}}{2\,(25\ \text{N-s}^2/\text{m}^2)(0.6\ \text{m/s}^2)(0.985)}$$

$$= \frac{1000}{2(25)(0.6)(1)}$$

$$= \frac{1000}{30}$$

$$x_1 = 33\ \text{m} \qquad \text{OK}$$

△ △ △ △ △ △ △ △ △ △ △ △ △

Check the size of the result

In cases where you have experience with the quantities you are calculating, judge whether the size of the result is reasonable. If the mass of a car comes out 2 kg, it had better be a toy car.

Check the exponent of the result more carefully than anything else. If the correct result is 3.03×10^3 m, or 3030 m, then writing 3.02×10^3 m by mistake gives a result of 3020 m, which is wrong by 10 m. Writing 3.03×10^2 m gives a result of 303 m, which is wrong by 2727 m.

Reread the problem

Go back to the original statement of the problem and check that you have done the problem that is asked for. You have already done this once before, after you put all of the given information down on your paper, but after you have worked the whole problem and have seen the result, you understand the problem better. Check again.

Mistakes

The reasons for working a problem clearly, step by step, are, first, to help in thinking about the problem and working toward a solution, second, to be able to understand the problem when you come back to it much later, and, third, to make it possible to find mistakes.

Mistakes are not completely avoidable. Only a small number of people who do calculations work without making mistakes. The methods we have been discussing will help you make fewer mistakes and find and correct them.

Once you make a mistake, it is not easy to find. You know that you can read your own essays several times and still not find simple mistakes in spelling and grammar. Our eyes see the right word even when the wrong word is on the page. Finding our own mistakes is certainly no easier with equations. That is why it is better to redo the solution a different way than to go over it again and again looking for mistakes.

A wrong assumption that is not written down or a wrong step done in your head is a dangerous thing. A mistake in a step that is not written down will never be found (except by your instructor or supervisor). If you write every step of a solution, put in every condition explicitly, and keep nothing in your head, you at least give yourself a chance to find mistakes.

Everyone goes up a few wrong paths at the beginning of the solution of any problem. If you make no mistakes of this kind, you are not doing hard enough problems. When you find the right path, solve the problem, recopy it, and hand it in, the wrong paths will not show. Wrong paths are not in the same category as mistakes in calculation that are caused by not following the rules of algebra or by not working in an orderly way.

Exercises

5–1. A hot-air balloon 35 m above a chained barking dog on the ground is moving toward the north-east while it is ascending at 3 m/s. Its total velocity is 10 m/s.

a. Find the balloon's horizontal velocity.

b. If the balloon is moving at constant velocity, find how far it is from the barking dog after 10 minutes.

5–2. In a molecule made up of two atoms, one of mass M_A and the other of mass M_B, the reduced mass μ is defined by

$$\frac{1}{\mu} = \frac{1}{M_A} + \frac{1}{M_B} .$$

If the mass of atom *A* is *R* times larger than the mass of atom *B*, solve for the reduced mass in terms of M_A and *R*.

5–3 A student at another school got this equation for a result

$$v = 2\pi aR$$

where *v* is a velocity, *a* is an acceleration, and *R* is a distance.

a. Is her result correct?

b. What mistake did she probably make?

5–4 One mole of any ideal gas contains 6.02×10^{23} molecules and occupies a volume of 0.0224 cubic meters at STP. The molecular weight of nitrogen (N_2) is 28. and the mass of one atomic mass unit is 1.67 yg. (1 yactogram = 10^{-24} g.)

a. Use these data to estimate, without using a calculator, the density at STP of nitrogen gas in kilograms per cubic meter.

b. Do the calculation with a calculator.

c. Calculate the percentage difference between your approximate and exact answers.

6

Presenting the Results

While you are solving a problem, it belongs only to you. You can try new approaches, go up wrong paths, redraw diagrams, make mistakes, and recopy the whole problem as many times as necessary.

When you make your results public by presenting them to others, you are staking your reputation on them. In college and afterward you will be evaluated on the correctness of your solution and on the clarity of your presentation.

Presenting your results can be as simple as drawing a box around your answer or as complicated as writing a scientific paper. The goal of presenting results is to tell someone else (or yourself a year later) what problem you solved, what answer you calculated, and how you arrived at that answer.

Present the problem and your solution clearly and completely enough that the reader can easily understand what you have done. Brilliant solutions badly presented are mostly ignored.

Numerical results

Put in known numbers

After you have written the result in symbols, put in the known numbers, all of which come from the data equations at the beginning of your description of the problem or from your calculation of intermediate values. To make it clear which symbol each number is replacing, write the numbers in the same relative positions as the symbols they replace. An equation containing numbers alone with no symbol equation above it is difficult to understand even a few minutes after it is written.

▽ ▽ ▽ ▽ ▽ ▽ ▽ ▽ ▽ ▽ ▽ ▽ ▽

To find x.

For motion under constant acceleration

$$x = x_0 + v_0 t + \frac{1}{2} a t^2$$

$$x_1 = 0 + 0 + \frac{1}{2} a_x t_1^2$$

$$= \frac{1}{2} a_x t_1^2$$

$$= \frac{1}{2} (0.6 \text{ m/s}^2)(10.02 \text{ s})^2$$

$$= 30.1 \text{ m}$$

△ △ △ △ △ △ △ △ △ △ △ △ △

Use significant figures to indicate accuracy

Use approximately the same number of significant figures for the result as the number of significant figures used in the original data for the problem. If the time is t=3.1 s and the distance is d=45 m, the velocity is v=14.5 m/s, not 14.516129032 m/s.

Since it is easy to calculate with high accuracy, use, during the intermediate calculations, two more significant figures than the original data contained. If you need to look up data that were not given in the statement of the problem, two extra significant figures keep you safe. If a radius is 45 m and you need to use the value of π, use π=3.142 to avoid unnecessary inaccuracy. The final result, however, should contain approximately the number of significant figures that the data contained.

In measurements and calculated results, zero serves two purposes. one is its usual purpose as a placeholder and the other is as an indicator of the accuracy of your result. If you report a result as 30.1 m, you are saying that your estimate of the accuracy of your measurement or calculation is about ±0.05 m, that is, plus or minus about half of the last significant figure. A result of 17 joules means you believe the answer is between 16.5 and 17.5 joules. If your result is 340 m, but your accuracy is no better than 5 m, write the result as 3.4×10^2 m.

Include units

After the numbers (or with the numbers) come the units.

▽ ▽ ▽ ▽ ▽ ▽ ▽ ▽ ▽ ▽ ▽ ▽ ▽

$$
\begin{aligned}
x_1 &= \frac{1}{2} a_x t_1^2 \\
&= \frac{1}{2}(0.6 \text{ m/s}^2)(10.02 \text{ s})^2 \\
&= 30.1 \text{ m}
\end{aligned}
$$

△ △ △ △ △ △ △ △ △ △ △ △ △

A numerical result without a unit is meaningless.

Use engineering prefixes

Large and small numbers are usually easier to understand if they are written with prefixes before their units. $4.5\text{x}10^{-5}$ meters becomes 45 µm. $67\text{x}10^9$ joules becomes 67 GJ. It is customary to use only those prefixes that step up and down by factors of 1000: milli, micro, nano, pico, fempto, atto, zepto, yacto (m, µ, n, p, f, a, z, y) going down, and kilo, mega, giga, tera, peta, exa, zetta, yotta (k, M, G, T, P, E, Z, Y), going up.

Put the result in a sentence

You have worked hard to describe the problem exactly and to solve it correctly. Now describe your result exactly. Write a sentence that exactly defines the variable you have calculated and that gives the numerical result and units.

▽ ▽ ▽ ▽ ▽ ▽ ▽ ▽ ▽ ▽ ▽ ▽ ▽

The distance the plane has traveled along the runway when it just lifts off is x_1=30.1 meters.

△ △ △ △ △ △ △ △ △ △ △ △ △

This finishes the problem with style and helps you and the person reading your work to remember just what it was that you calculated. There are even times when writing this final sentence suddenly makes you realize that the quantity you calculated was not the quantity you wanted.

Decorate your results

Present your results so that the reader can find them instantly. In a five page calculation the results may take up only one line, but the person reading the calculation wants to see that one line first. Make it easy to find. Write the result slightly larger than the rest of the calculation and put a box around it. The result includes a symbol, an equals sign, a number, its unit, and a sentence describing the result in English.

▽ ▽ ▽ ▽ ▽ ▽ ▽ ▽ ▽ ▽ ▽ ▽ ▽

The distance the plane has traveled along the runway when it just lifts off is x_1=30.1 meters.

△ △ △ △ △ △ △ △ △ △ △ △ △

Graphs

Graphical results are easier to absorb and think about than equations, tables, or numbers. Computers have made graphing easy and fast, but it is still up to you to design understandable graphs.

Title

Every graph needs a title. Although *you* know exactly what you plotted, the person reading the graph usually does not, and it is the title that tells her. The title—often in a box on the graph—describes what is plotted and the conditions under which the plot is true.

▽ ▽ ▽ ▽ ▽ ▽ ▽ ▽ ▽ ▽ ▽ ▽ ▽

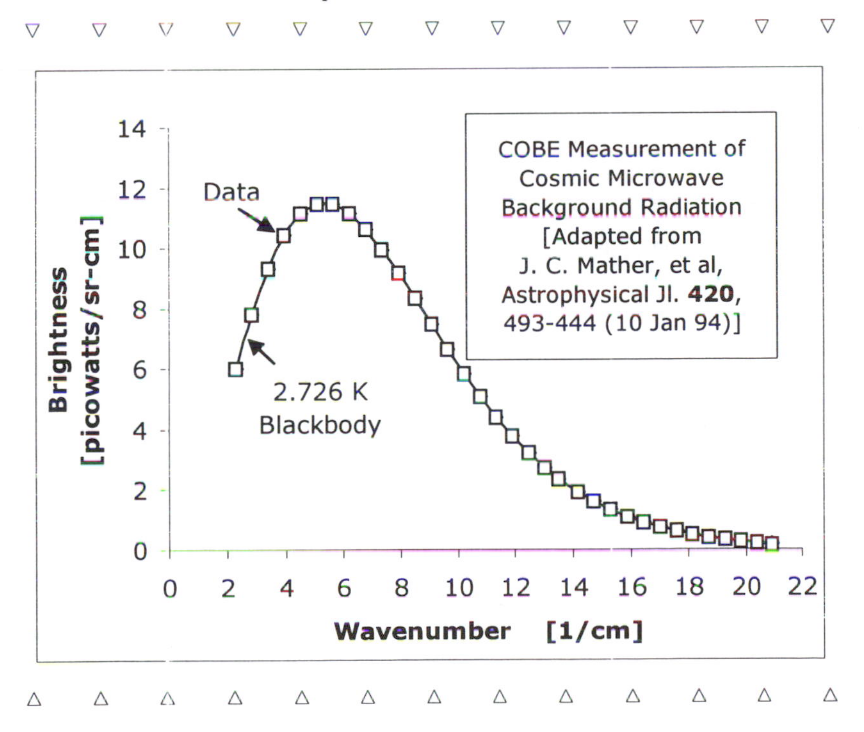

△ △ △ △ △ △ △ △ △ △ △ △ △

Axis labels

The x (horizontal or left-right) axis is usually used for the independent variable. and the y (vertical or up-down) axis is used to plot the dependent variable. For a given value of x, one can read a value of y from the graph.

When graphing something as a function of time, the time is the independent variable, and the time axis runs from left to right.

Label each axis with the full name and unit of its variable. The full name is better than the symbol because it is easier to understand. A graph without these labels is almost impossible to interpret.

Scale divisions

Both the x and y axes need major numbered scale divisions and minor unnumbered divisions. Choose these divisions so that the reader can find the value of a point on the axis between two of the marked divisions. Usually the distance between the divisions is a power of 10 times 1, 2, or 5. Major divisions can be placed at 0.1, 0.2, 0.3, 0.4 ..., with minor divisions spaced by 0.01 or 0.02 or 0.05. Or major divisions can be placed at 10, 20, 40, 60 ..., with minor divisions spaced by 1, 2, or 5, and so forth. These rules are the same that you used when dividing axes on your diagrams at the beginning of the problem.

If a graph gives detailed information that readers may want to convert into numbers, put scale divisions on the top and right hand borders of the graph as well as along the axes. The reader can then lay a ruler between the left and right divisions and read accurate values from the graph.

Points and lines

The results of a calculation are usually plotted as a continuous line with no symbol marking the individual points. Although you calculated the value of y only at some values of x, join the calculated points by straight lines to form a curve.

Experimentally measured points are usually plotted as individual points using symbols such as □ and △.

Straight lines

The easiest graph to understand is a straight line. If a graph is not a straight line, sometimes it can be replotted using different variables to make it a straight line. For instance, $y=x^2$ is not a straight line when y is plotted against x, but becomes a straight line if y is plotted against x^2.

Logarithmic axes

If either x or y covers a wide range (many powers of 10) a graph that shows the large values will give very little detail about the small values. To get more information from the graph make the scale of the wide-range variable logarithmic.

A logarithmic scale means that instead of the position of a value along the axis being proportional to the value itself, the position is proportional to the logarithm of the value. If 1 inch along the axis corresponds to 1, 2 inches corresponds to 100, 3 inches corresponds to 1000, etc. Although the data between 1 and 10 are hard to see on a linear graph that goes up to 1000, on a log graph they are given just as much room as the data between 100 and 1000.

On a log graph, since the distance between 1 and 10 is the same as the distance between 10 and 100, the distance between 1 and 2 is not the same as the distance between 2 and 3. Log graphs have evenly spaced divisions at .001, .01, 0.1, 1, 10, 100, 1000, ... and unevenly spaced divisions at 0.002, 0.003, 0.004, ...

On a graph with a logarithmic y axis, the curve $y=e^x$ is a straight line, so that pure exponential functions are easy to recognize on logarithmic graphs.

Rewriting

Sometimes it pays to rewrite the whole solution. After you have the answer, the problem becomes much clearer than when you began. Now you can write the whole solution in a clear logical way that will show the instructor the best that you can do and will also be understandable later when you are studying for the final exam.

Avoid rewriting exam problems. On an exam, rewriting takes extra time and is confusing to the grader. The time you save by not rewriting you can spend doing the solution neatly the first time. Practice this on homework problems by doing them neatly enough so that most of them do not need rewriting.

Reports and Publications

If you turn your solution of a problem into a report or publication, you will have to follow the rules of the organization or journal to which you submit it. Everything will have to be written in complete English sentences and the ratio of explanation to calculation will go way up. Often, your result and its interpretation will follow directly after your statement of the problem, and your calculations will come later, or, sometimes, be omitted entirely. Each equation will need a number. Figures will have to be separated from the text and each figure will need a caption. The report or publication will need an abstract and a list of references to other published work.

Exercises

6–1. The acceleration of gravity at the surface of the earth—caused by the earth's mass—is 9.8 m/s^2. At the equator, the downward acceleration caused by the earth's rotation is v^2/R, where v is the velocity of rotation of a point on the equator, and R is the earth's radius. If the earth's rotation were speeded up until the rotational acceleration equaled the gravitational acceleration, people on the equator would feel no gravitational force. Since the time it takes for one rotation of the earth is called a day, how many seconds would there be in a speeded-up earth day?

6–2. An elevator in the Sears Tower is passing the ground floor when a passenger starts her stopwatch. At 2 seconds the elevator is at the 3rd floor, at 8 seconds it is at the 12th floor and at 12 seconds it is at the 18th floor. The floors are each 4.0 m high.

a. Make an accurate scale drawing at a scale of 1:500 of the position of the elevator at each time.

b. Make a graph of the position of the elevator vs. time.

c. Connect the measured points with a straight line.

d. Extend the line to predict when the elevator will reach the 24th floor.

e. Use the slope of the line to find the velocity of the elevator.

6–3. An operational amplifier (opamp) amplifies its input voltage V_i to produce an output voltage V_o. The gain g ($= V_o/V_i$) depends on the frequency f with which the input voltage is changing. When $f = 10$ Hz (10 cycles per second), $g = 10$ million. As f increases, g goes down. Every time the frequency doubles, the gain is halved, that is, when $f = 20$ Hz, $g = 5\text{x}10^6$, and so forth.

a. Write an equation for the gain as a function of frequency $g(f)$.

b. Calculate the gain of the opamp for frequencies from 10 Hz to 1 MHz.

c. Plot the gain vs. frequency.

d. Plot the gain vs. frequency on a graph on which both the gain and the frequency scales are logarithmic.

7

Can't Solve It

You have done an accurate drawing, defined symbols for all the knowns and unknowns, converted units, and written the preliminary equations, and you still don't know how to solve the problem. Now what?

Use thinking where it counts

When you are attacking a problem that you do not know how to do and you have finished the automatic activities, there are few substitutes for organized thinking. Many of the how-to-solve-it books listed at the end of this book discuss this kind of thinking. We give just a brief summary here.

Look for a similar problem

Look in the textbook or in the literature for a similar problem that has been solved. Seeing the solution of a similar problem can give you ideas on how to proceed. The library usually has a collection of introductory textbooks, all of which contain solved examples. Looking at the way material is presented in other textbooks can often help you to get through a difficult course. Sometimes, doing other problems in your textbook, especially the problems that come just before or after the assigned problem, can give you a hint and get you started.

Simplify

Make up a similar problem that is simpler than the problem that you want to solve. Working on a simpler problem can give you a start and can show you the structure of the more complicated problem.

Generalize

Sometimes you can make progress by generalizing the problem. Remove some of the constraints until the problem is very general. This is only useful in some cases, but when it works, it lets you solve a whole class of problems instead of one particular problem.

Put in numbers

If you have found equations that connect the known and unknown quantities in a problem but cannot combine the equations to find the answer, putting in numbers can help. You may be used to solving

$$3x - 5y = 5$$

$$7x + 2y = 3$$

for x and y, but may not be experienced with

$$ax - by = c$$

$$dx + ey = f \ .$$

Although the preferred way of solving a problem is to use symbols for the known and unknown quantities until the last step, by putting in the known numbers you may increase your chances of success with equations. If you have trouble at the beginning of a course, by all means, put in numbers. As you become more experienced, try to use symbols instead.

When a problem contains no known quantities at all, using simple numbers, 1, 10, etc., in place of the abstract quantities described in the problem can make the ideas in the problem easier to understand. Going 10 miles in one hour is easier to understand than is $v = d/t$. Once you straighten out the ideas using numbers, you may be able to solve the problem using symbols.

Look for unused data

If you have done part of a problem and cannot do the rest, compare the data given in the statement of the problem with the data you have used. The data that were given but not used may provide the clue to solving the remainder of the problem.

Try a ratio

If you seem not to have enough information to find a quantity that is asked for, try dividing it by a quantity of the same type. Divide a mass by a mass, a velocity by a velocity, and so on. Sometimes the same unknown quantity appears in the top and bottom of the right hand side of the ratio equation and cancels.

Put it aside

When you have described a problem on paper and understood it as well as you can but still cannot solve it, put it aside for a short time, or better, overnight. If you have worked hard enough on a problem and thought about it for a long enough time, sometimes the back of your mind can solve a problem that the front cannot. The larger a problem is, the better this works.

Go for a little help

After drawing a diagram, making a list of symbol definitions, writing the data equations, converting the units, writing the preliminary equations, looking for a similar problem, simplifying the problem, and generalizing the problem, you still do not know how to solve it, go for a little help.

Find a teacher, a learning center tutor, or a colleague who is willing to act as a consultant. It is not necessary that they know how to do the problem, but it is better if they have experience doing problems.

Asking someone for help with a problem is good practice for what you will be doing as a scientist or engineer. Asking for help does not show weakness. It shows that you are willing to learn, that you realize you cannot know everything, and that you know the most efficient way to get your work done. Since asking for help requires you to explain your problem to someone else, it forces you to state the problem clearly. In fact, describing a problem in a simple understandable way to someone who knows nothing about the subject can unlock your creativity in finding a solution.

How to ask for help

By going to a teacher for help you will learn how to present an unfinished problem to a consultant in an efficient way. You have only a few seconds in which to introduce yourself by name, state the problem, show your preparations, and ask for advice. As soon as you get a new idea, leave, and try to finish the problem by yourself. Going back for help a second time shows commitment and tenacity.

Introduce yourself by name. Saying, "Hello, I'm Edward Emory," is not only polite, but also avoids the awkwardness some teachers feel when having to ask you your name. Rather than saying, "I can't do problem seven," show the original statement of the problem, show your drawing, definition list, and other preparations, and discuss your ideas.

As soon as your consultant provides a single new idea that you think may help, leave. Go away and work on the problem again yourself. If there is still trouble, go again for a little help. As soon as a single useful idea surfaces, go away and work by yourself.

If you allow someone else to solve your problem, that problem is spoiled forever. You will never understand every step until you have invented and thought about every step yourself. Those steps that someone else does are just the ones you will not understand on the next problem or on the exam. Get help, but do the problem yourself.

8

Spreadsheets

Computer spreadsheets are a fast and easy way to do calculations. A spreadsheet can record and analyze data, calculate functions, and make graphs in addition to its traditional strengths of keeping lists of addresses, preparing budgets, maintaining a checking account, or running a small business.

A spreadsheet encourages breaking a problem into parts and writing the parts separately. It can be made self documenting, displaying separately the definition, name, value, and unit for each variable. You can comment on each step, making it easy to understand the calculation and to find errors.

Since computer spreadsheets have all of the functions that computer languages have, such as *sin*, *exp*, *rnd*, and *abs*, any equation that can be written in a computer language can be written in a spreadsheet cell.

You can set up and debug a spreadsheet in far less time than it would take to write a computer program. Much of a well-structured program consists of formatted input statements, formatted output statements, and loops. A computer spreadsheet does these things automatically, leaving you free to work on the structure of the problem and the equations.

A spreadsheet is an array of numbered rows (across) and lettered columns (down) of cells. Each cell can hold a number, a name, a comment, or an equation. To tell the spreadsheet that an entry is an equation, begin the entry with an arithmetic operator like = or + . When you type an equation into a cell and press <return> the spreadsheet calculates the equation and shows the result in the same cell. The equation is still there and can be seen by selecting that cell again.

An equation can refer to other cells by using either the locations or the names of those cells. Thus = A1*A2 means multiply the number in cell A1 by the number in cell A2 . If you later change the number in A1, the equation will use the new number. Even more elegantly, you can give cells names, and use those names in calculations. If you name cell A1 *velocity* and cell A2 *time*, then a cell with the equation =*velocity* ∗ *time* will calculate the distance traveled.

In pen and paper calculations it is best to do the solution algebraically and to insert the data numbers in only the last equations. In a spreadsheet calculation, you can see both the algebraic equations and the numerical value of those equations all the way through the calculation. This helps in thinking about the problem and in avoiding mistakes.

Spreadsheets can make a graph by plotting the numbers in one column against the numbers in another column. If you make changes in either column, the graph shows those changes immediately.

A powerful way to analyze experimental data is by fitting an equation to the data using a spreadsheet. Adjust the parameters in the equation until the calculated curve agrees with the measured points.

Spreadsheets are a natural way to set up matrix calculations, especially those in which successive matrices act on a column vector, such as in geometrical optics and polarization calculations.

Computers have made spreadsheets into powerful calculational tools. The computer can calculate the left hand side of each equation on the spreadsheet as long as the value of every quantity on the right hand side exists somewhere on the spreadsheet. A computer spreadsheet not only helps to organize and document the solution to a problem but does all of the arithmetic correctly as well.

Calculating a single value

For calculations that have only one value for each variable, put each variable on a separate line with its definition, its name, its value or equation, and its unit.

The next two sections are an example of this kind of problem and its spreadsheet solution.

▽ ▽ ▽ ▽ ▽ ▽ ▽ ▽ ▽ ▽ ▽ ▽ ▽

Example 5: Solar rocks

A solar collector absorbs 1 kilowatt of sunlight for a period of 6 hours. The collected energy is used to heat a metric ton (1000 kg) of rocks whose specific heat capacity is half that of water. How much does the collected solar energy increase the temperature of the rocks?

Solution for Example 5

Here are the definitions and equations as you would enter them in a spreadsheet. (You can display the equations rather than their results by choosing the menu item **Display Equations**.)

	A	B	C	D
1	**Solar Rocks**			

2	D. Scarl			
3	31047			
4	**Definitions and Data:**			
5	power absorbed	PkW	1	kW
6	time period	th	6	hours
7	mass of rocks	mt	1	ton
8	ratio of specific heat capacity of rocks to that of water	s	0.5	
9	**Constants:**			
10	specific heat capacity of water (Handbook of Chemistry and Physics, 52nd Ed. Chemical Rubber Publishing Co. 1971, pg D-128)	cw	4.18E+03	joules/(kg-K)
11	**Unit Conversion:**			
12	power absorbed	PW	=PkW*1000	watts
13	time period	ts	=th*3600	sec
14	mass of rocks	mkg	=mt*1000	kg
15	**Preliminary Equations:**			
16	energy absorbed	Q	=PW*ts	joules
17	specific heat capacity of rocks	cr	=s*cw	joules/(kg-K)
18	Result:			
19	temperature increase of rocks	Tr	=Q/(cr*mkg)	K

Here are the results of the equations as calculated by the spreadsheet. Now cells C12 to C19 display the calculated results in place of the equations.

	A	B	C	D
1	**Solar Rocks**			
2	D. Scarl			
3	23 February 1993			
4	**Definitions and Data:**			
5	power absorbed	PkW	1	kW
6	time period	th	6	hours
7	mass of rocks	mt	1	ton
8	ratio of specific heat capacity of rocks to that of water	s	0.50	

9	**Constants:**			
10	specific heat capacity of water (Handbook of Chemistry and Physics, 52nd Ed. Chemical Rubber Publishing Co. 1971, pg D-128)	cw	4.18E+03	joules/(kg-K)
11	**Unit Conversion:**			
12	power absorbed	PW	1000	watts
13	time period	ts	21600	sec
14	mass of rocks	mkg	1000	kg
15	**Preliminary Equations:**			
16	energy absorbed	Q	21.6E+06	joules
17	specific heat capacity of rocks	cr	2090	joules/(kg-K)
18	**Result:**			
19	temperature increase of rocks	Tr	10.3	K

△ △ △ △ △ △ △ △ △ △ △ △ △

Inputs

The first section of the spreadsheet contains all of the inputs. These are the data equations of the problem. Each input equation is on a separate line, with the definition in column A, the name in column B, the value in column C, and the unit in column D.

▽ ▽ ▽ ▽ ▽ ▽ ▽ ▽ ▽ ▽ ▽ ▽ ▽

	A	B	C	D
4	**Definitions and Data:**			
5	power absorbed	PkW	1	kW
6	time period	th	6	hours
7	mass of rocks	mt	1	ton
8	ratio of specific heat capacity of rocks to that of water	s	0.5	

△ △ △ △ △ △ △ △ △ △ △ △ △

Use the menu to tell the spreadsheet to give cells C5, C6, and C7 the names you wrote in cells B5, B6, and B7, that is, cell C5 is named *PkW*, cell C6 is named *th*, and cell C7 is named *mt*. Then you can use these names anywhere on the spreadsheet.

Constants

Constants are quantities that may not be given in the statement of the problem but are needed for its solution. They can be properties of materials, like density, or fundamental constants, like the acceleration of gravity. Put each constant on a separate line with its definition, its symbol, its value, and its unit.

▽ ▽ ▽ ▽ ▽ ▽ ▽ ▽ ▽ ▽ ▽ ▽ ▽

	A	B	C	D
9	**Constants:**			
10	specific heat capacity of water (Handbook of Chemistry and Physics, 52nd Ed. Chemical Rubber Publishing Co. 1971, pg D-128)	cw	4.18E+03	joules/(kg-K)

△ △ △ △ △ △ △ △ △ △ △ △ △

Unit conversion

These spreadsheet rows convert the given units into standard SI units.

▽ ▽ ▽ ▽ ▽ ▽ ▽ ▽ ▽ ▽ ▽ ▽ ▽

	A	B	C	D
11	**Unit Conversion:**			
12	power absorbed	PW	=PkW*1000	watts
13	time period	ts	=th*3600	sec
14	mass of rocks	mkg	=mt*1000	kg

△ △ △ △ △ △ △ △ △ △ △ △ △

When the unit of a quantity changes on a spreadsheet, its name must change too. Cells C12 calculates *PW*, the power in watts, using *PkW*, the power in kilowatts, which is the name that you gave to Cell C5.

Calculation

The steps of the calculation can be done row by row down the spreadsheet so that the definitions, the data equations, and the calculation have the same format.

V

	A	B	C	D
15	**Preliminary Equations:**			
16	energy absorbed	Q	=PW*ts	joules
17	specific heat capacity of rocks	cr	=s*cw	joules/(kg-K)

△ △ △ △ △ △ △ △ △ △ △ △ △

Results

The last few rows of the spreadsheet present the results. Again, each result has its definition in the first column, its symbol in the second, its value in the third, and its unit in the fourth.

You type this.

▽ ▽ ▽ ▽ ▽ ▽ ▽ ▽ ▽ ▽ ▽ ▽ ▽

	A	B	C	D
18	**Result:**			
19	temperature increase of rocks	Tr	=Q/(cr*mkg)	K

The spreadsheet displays this.

	A	B	C	D
18	**Result:**			
19	temperature increase of rocks	Tr	10.3	K

△ △ △ △ △ △ △ △ △ △ △ △ △

Plotting data and calculating functions

The next example shows three separate uses of a spreadsheet. The first is a simple plot of experimental data. The second is the plot of a function whose equation is known. The third is the adjustment of constants in the equation to make the plot of the equation fit the plotted data.

▽ ▽ ▽ ▽ ▽ ▽ ▽ ▽ ▽ ▽ ▽ ▽ ▽

Example 6: Optical detector power

A compact disc player reads a CD by shining a laser beam through a lens onto the surface of the CD. The CD reflects the laser light back through the lens onto a photodiode that converts the light power P to an electrical current that is proportional to the light power. If x, the distance between the surface of the CD and the lens, is too small or too large, the light makes an out-of-focus spot at the photodiode, reducing the detected power. Here are data recorded by moving the lens toward and away from the surface of the CD while recording the power falling on the photodiode.

x mm	P mW
0.50	0.05
0.60	0.35
0.70	0.55
0.80	0.60
0.90	0.55
1.00	0.35
1.10	0.05

a. Plot the detected power P as a function of the distance x between the CD and the lens.

b. The data can be described by the equation

$$P = P_{max}(1-a(x-x_0)^2)$$

where x_0 is the optimum distance between the CD and the lens, P_{max} is the power detected by the photodiode at the optimum distance, and a is an adjustable parameter that controls the width of the function. (This equation only describes the data over a limited range, since P can go negative, but the measured power cannot.)

Plot $P(x)$ and find numbers for P_{max}, x_0, and a that give a reasonable fit to the data.

Solution for Example 6

	A	B	C	D
1	**CD Detector Power**			
2	D. Scarl			
3	24 Jun 94			
4				
5	distance from CD to lens	x		mm
6	measured light power at detector	Pmeas		W
7	calculated light power at detector	P		mW
8	maximum light power	Pmax	0.6	mW
9	distance from CD to lens at maximum power	xzero	0.8	mm
10	width constant	a	10	mW/mm2
11				
12			P =	
13	x	Pmeas	Pmax*(1–a*(x–xzero)^2)	
14	mm	mW	mW	
15	0.50	0.05	0.060	
16	0.55		0.225	
17	0.60	0.35	0.360	
18	0.65		0.465	
19	0.70	0.55	0.540	
20	0.75		0.585	
21	0.80	0.60	0.600	
22	0.85		0.585	
23	0.90	0.55	0.540	
24	0.95		0.465	
25	1.00	0.35	0.360	
26	1.05		0.225	
27	1.10	0.05	0.060	

a. Plot the measured power against the distance between the CD surface and the lens.

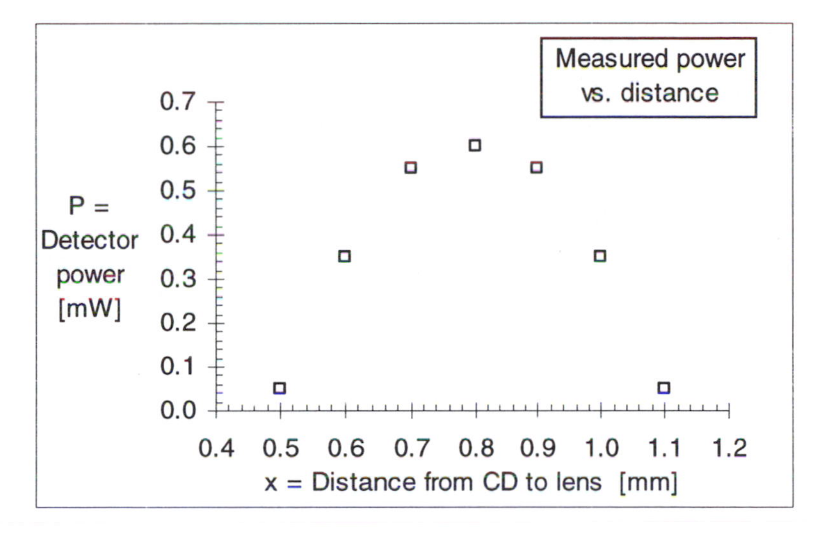

b. Plot $P = P_{max}(1 - a(x - x_0)^2)$ and adjust P_{max}, x_{zero}, and a to fit the data.

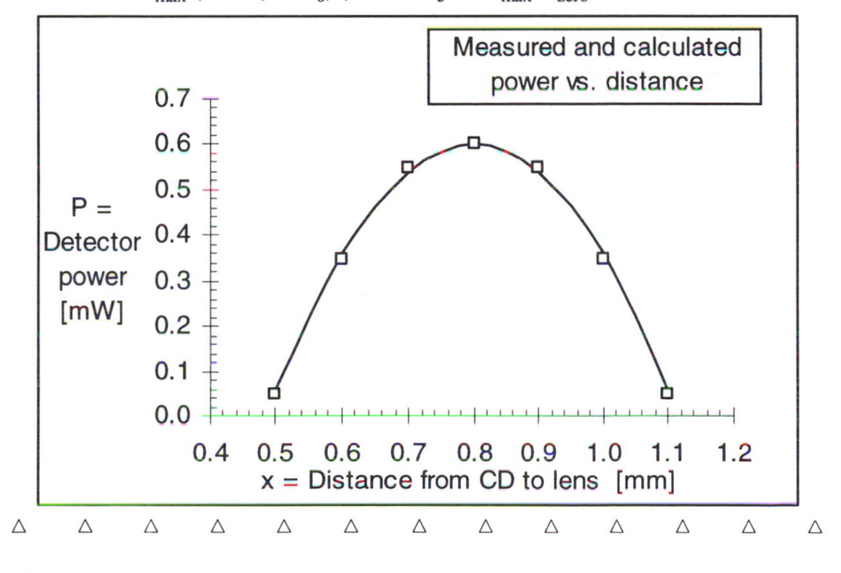

△ △ △ △ △ △ △ △ △ △ △ △ △

Graphing data

If you tell a spreadsheet that the numbers in its first column are the values of the independent variable, it can plot the values in the second column against those in the first column.

▽ ▽ ▽ ▽ ▽ ▽ ▽ ▽ ▽ ▽ ▽ ▽ ▽

	A	B
13	x	Pmeas
14	mm	mW
15	0.50	0.05
16	0.55	
17	0.60	0.35
18	0.65	
19	0.70	0.55
20	0.75	
21	0.80	0.60
22	0.85	
23	0.90	0.55
24	0.95	
25	1.00	0.35
26	1.05	
27	1.10	0.05

△ △ △ △ △ △ △ △ △ △ △ △ △

In Example 6 the numbers in the first column are the distance x from the CD surface to the lens and the numbers in the second column are the power P_{meas} detected by the photodiode. The extra values of x (0.55, 0.65, ...) are not used in this plot; they will be used to make a smoother curve in the next part of the problem.

Select lines 13 to 27 and columns A and B and choose **graph** from the menu to make a graph of P_{meas} vs. x.

▽ ▽ ▽ ▽ ▽ ▽ ▽ ▽ ▽ ▽ ▽ ▽ ▽

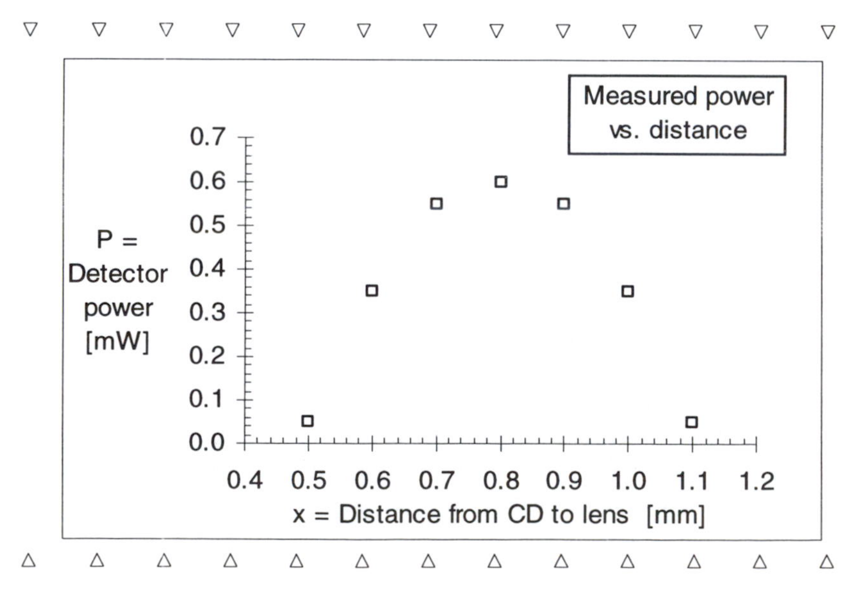

△ △ △ △ △ △ △ △ △ △ △ △ △

Graphing a function

To plot the graph of a function put the values of the independent variable in the first column and the equation for the dependent variable in the next available column.

▽ ▽ ▽ ▽ ▽ ▽ ▽ ▽ ▽ ▽ ▽ ▽ ▽

	A	B	C	D
8	maximum light power	Pmax	0.6	mW
9	distance from CD to lens at maximum power	xzero	0.8	mm
10	width constant	a	10	mW/mm2
11				
12			P =	
13	x	Pmeas	Pmax*(1–a*(x–xzero)^2)	
14	mm	mW	mW	
15	0.50	0.05	0.060	
16	0.55		0.225	
17	0.60	0.35	0.360	
18	0.65		0.465	
19	0.70	0.55	0.540	
20	0.75		0.585	
21	0.80	0.60	0.600	
22	0.85		0.585	
23	0.90	0.55	0.540	
24	0.95		0.465	
25	1.00	0.35	0.360	
26	1.05		0.225	
27	1.10	0.05	0.060	

△ △ △ △ △ △ △ △ △ △ △ △ △

Rows 8, 9, and 10 define the constants *Pmax*, *xzero*, and *a*. If the constants were given in the statement of the problem, they would go here. Since Example 6 does not give the values of these constants, put in any numbers and adjust them later to fit the curve to the experimental data.

After the three rows of constants column A contains all of the values of the independent variable. Column B contains the experimental data.

Column C is the calculation of the diode power *P*. Cell C13 just displays the equation for *P* that will be calculated in all the rows below. The equation in cell C13 is not calculated because the first symbol in the cell is not = or +.

Cells C15 to C27 do the calculation of the power; each contains the same equation.

▽ ▽ ▽ ▽ ▽ ▽ ▽ ▽ ▽ ▽ ▽ ▽ ▽

	A	B	C	D
15	0.50	0.05	=Pmax*(1–a*(x–xzero)^2)	
16	0.55		=Pmax*(1–a*(x–xzero)^2)	
17	0.60	0.35	=Pmax*(1–a*(x–xzero)^2)	

△ △ △ △ △ △ △ △ △ △ △ △ △

The spreadsheet does not show this equation. Instead it shows its value. When $x = 0.50$, $P = 0.060$, and so forth. (To be able to use the cell name x in column C, give the name x to all the cells from A15 to A27.)

To plot the data and the function P on the same graph, select columns A, B, and C, and rows 13 to 27. Choose **graph** from the menu, and both the data and the function will be plotted against x. Use the menus or dialog boxes to format the data points as squares and the function as a line. Add labels to the horizontal and vertical axes, adjust the scale divisions, and add a title box to make a complete graph. This graph is shown above in Solution for Example 6.

Fitting a function to data

Change the values of *Pmax, xzero*, and a in cells C8, C9, and C10 and watch how the curve changes. Choose values of *Pmax*, *xzero*, and a that bring the curve as close as possible to the data, so that there are about the same number of data points above and below the curve. In the Solution to Example 6, *Pmax*, *xzero*, and a have been chosen to be 0.6, 0.8, and 10 to create a curve that fits the data.

Numerical differentiation

You can find the slope of a curve at a point by subtracting the y values of the two points that surround it, subtracting the x values of the same two points, and dividing the difference in y values by the difference in x values. On a spreadsheet the equations look like this.

	A	B	C
1	x	y	dy/dx
2	0.50	0.060	
3	0.55	0.225	=(B4 – B2)/(A4 – A2)
4	0.60	0.360	=(B5 – B3)/(A5 – A3)
5	0.65	0.465	=(B6 – B4)/(A6 – A4)
6	0.70	0.540	=(B7 – B5)/(A7 – A5)
7	0.75	0.585	=(B8 – B6)/(A8 – A6)
8	0.80	0.600	=(B9 – B7)/(A9 – A7)
9	0.85	0.585	

The results look like this.

	A	B	C
1	x	y	dy/dx
2	0.50	0.060	
3	0.55	0.225	3
4	0.60	0.360	2.4
5	0.65	0.465	1.8
6	0.70	0.540	1.2
7	0.75	0.585	0.6
8	0.80	0.600	0
9	0.85	0.585	

Numerical integration

You can calculate an integral numerically by calculating the value of the function under the integral for many values of the variable of integration and then weighting and adding the calculated values. The standard ways to do this are described in books on numerical methods.

To do numerical integration on a spreadsheet put the values of the integration variable in the first column and calculate the function in the usual way. Include a column for the weights and a final column for the product of the weights times the values of the function. To calculate the integral, sum the values in the final column.

Other operations

A spreadsheet can sort data alphabetically or numerically. Sorting is useful for statistical calculations such as finding medians and percentiles. A spreadsheet can calculate the mean and standard deviation of a set of data and do correlation and regression calculations.

Fitting a curve to a set of data can be done by eye as in Example 6, or the spreadsheet can do it automatically.

Since a spreadsheet is already an array, you would expect it to excel in doing matrix calculations. A spreadsheet can add vectors and matrices, can multiply a vector times a vector, a matrix times a vector, or a matrix times a matrix. It can also invert a matrix and find its trace, determinant, and eigenvalues.

Organizing a spreadsheet

The rules for setting up a spreadsheet to solve a problem are almost the same as the rules discussed in the previous chapters for setting up any problem. Break the spreadsheet up into parts. First write the data given in the original statement of the problem, then constants and other outside values. Write the preliminary equations, the science equations, and finally, the results.

If an algebraic expression is complicated, break it into parts, calculate each part in a separate cell or column, and then combine the parts to get the result. This makes the equation in each cell easy to write, understand, and debug.

Avoid numbers

A good rule for spreadsheet equations—and for equations in any computer program—is to avoid numbers in the body of the calculation. If you put a number into an equation in the main part of the calculation, it will be copied into many other cells and will be hard to find and change. Put each number into a named cell and then use the name in the body of the calculation. When you change the number in the named cell, it will automatically change everywhere in the calculation.

Giving cells names

When you define a quantity in one row of a spreadsheet you provide a definition, a name, a number, and a unit.

▽ ▽ ▽ ▽ ▽ ▽ ▽ ▽ ▽ ▽ ▽ ▽ ▽

	A	B	C	D
8	maximum light power	Pmax	0.6	mW

△ △ △ △ △ △ △ △ △ △ △ △ △

The spreadsheet, though, does not know that the number in cell C8 has a name unless you specifically tell it. You can do this in two ways. One is to select cell C8 and ask the spreadsheet to name it *Pmax*. The other is to select cells B8 and C8 and tell the spreadsheet to use whatever is in the left hand cell as the name of the right hand cell. By selecting more than one row in columns B and C you can use a single command to name a group of cells.

Choosing cell names

Naming cells in a computer spreadsheet is slightly different from naming quantities in a calculation on paper. Some spreadsheets do not allow subscripts or superscripts, so names such as v_{x0} have to be written as *vx0* or, better, *vxzero*.

Since spreadsheet columns are designated by letters and rows by numbers, the location of a cell is one or two (the column after Z is AA) letters followed by a number. If you give a cell a name that is one or two letters and a single number (*vx0*), it will look like a location rather than a name and will cause trouble. Giving cells longer names, such as *vxzero*, avoids this problem and also helps you to remember what the names mean.

Editing a spreadsheet

To write almost the same equation in many cells in a column, write the equation in the top cell, select that cell and the rest of the column, and choose **Fill Down** from the menu. The spreadsheet will copy the equation into all the cells below. The spreadsheet knows that the equation needs to change slightly in each row. For instance, if column A of the spreadsheet contains values of x from 0 to 1 in steps of 0.1, and you want to plot $1/x^2$, the first cell of column B contains =1/A1^2, the second cell contains =1/A2^2, the third cell, =1/A3^2, and so forth. All you need to write is the equation in the top cell, =1/A1^2. When it copies this equation into the next cell down, the computer automatically moves the reference one cell down so that now the equation reads =1/A2^2. With a single command, you can fill an entire column with the same equation, each row referring to the proper value in the other columns. (If you have given the name x to the cells in column A, all of the equations in column B will simply be = 1/x^2.)

If an equation uses a number from a cell in your data equations, say C8, you do not want the cell location to change when you copy the equation to the next row. You can avoid having the spreadsheet automatically put C9 in the next row by preceding the cell location by a special symbol, often *$*. A reference to cell C8 will not change to C9 when you copy it to the row below. (If you have given the name *Pmax* to cell C8 you avoid this problem altogether.)

The spreadsheet can automatically generate equally spaced values of a variable. This is useful for creating a column containing values of an independent variable that will be used to calculate a function. Type the starting value in the top cell, select that cell and the cells below it and choose **Data Series** from the menu. The spreadsheet will automatically fill the column with equally spaced numbers.

Choosing a type of graph

When asked to plot a graph, most business spreadsheets assume that the values in the first column are labels, like January, February, March, ..., and space them equally along the axis no matter what their value. In business-oriented spreadsheets, this is called the category axis because it contains the various categories, each of which has a value to be represented on the graph. The perpendicular axis is called the value axis.

A technical graph has to place the values of the independent variable along the horizontal axis at distances proportional to their numerical value. If the values of your independent variable are not equally spaced, but the spreadsheet plots them at equally spaced positions along the horizontal axis, the graph will not be correct. Choose a type of graph called **x-y plot** or **scatter plot** to tell the spreadsheet that the numbers in the first column are values of the independent variable and that you want them positioned along the horizontal axis at distances proportional to their numerical values.

Exercises

8–1. The recording area on a compact disc has an inside diameter of 1.7 inches and an outside diameter of 4.5 inches. Each recorded bit occupies an area of 1.8 μm by 1.8 μm. Music requires recording 16 bits 40,000 times each second. Calculate the maximum recording time for a compact disc.

8–2. The average distance from the earth to the sun, $1.4x10^8$ km, is called an astronomical unit AU. The distances of the planets from the sun are

Planet	Number	Distance
Mercury	1	0.4 AU
Venus	2	0.7 AU
Earth	3	1.0 AU
Mars	4	1.5 AU
Jupiter	6	5.2 AU
Saturn	7	9.5 AU
Uranus	8	19.2 AU

a. Use a spreadsheet to plot the planetary distances vs. planet number.
b. An empirical law that has no theoretical foundation (Bode's law) says that the distances from the planets to the sun are given by

$$d = 0.4 + (0.3)2^{(n-2)}$$

where d is the average distance from the planet to the sun in AU and n is the number of the planet. (For Mercury, the second term is zero.) Plot Bode's law on the same graph as the measured planetary distances.
c. It is believed that the asteroid belt is made from pieces of a planet that broke up or was never formed. From your plot estimate the location of the asteroid belt.

Problems

1. Satellite period. A satellite in a circular orbit 640 km above the surface of the earth travels at 7.54 km/s.
a. Write a heading, draw a diagram, and label it with all of the symbols for all of the quantities you will need in order to answer part d.
b. Make a list of definitions, symbols, values, and units for all of the quantities you will need to answer part d.
c. Make the preliminary unit conversion and geometric calculations that will help to solve part d.
d. How long does it take the satellite to make one trip around the earth?

2. Shortcut. Your college quadrangle is 85 meters long and 66 meters wide. When you are late for class you can walk at 7 miles per hour. You are at one corner of the quad and your class is at the diagonally opposite corner.
a. Write a heading, draw a diagram, and label it with all of the symbols for all of the quantities you will need in order to answer part d.
b. Make a list of definitions, symbols, values, and units for all of the quantities you will need to answer part d.
c. Make the preliminary unit conversion and geometric calculations that will help to solve part d.
d. How much time can you save by cutting across the quad rather than walking around the edge?

3. Distant lake. An airliner flying at a height of 11 000 m is directly above a small lake. From where you are standing, at the same height as the lake, the angle from the horizon to the airliner is 32°.
a. State the problem exactly by writing a heading, drawing a diagram, and labeling it with all of the symbols for all of the quantities in the statement of the problem.
b. Make a list of definitions, symbols, values, and units for all of the quantities in the statement of the problem.
c. Add to your diagram and list of definitions the quantities you will need to calculate to find the answer to part e.
d. Make the preliminary unit conversion and geometric calculations that will help to solve part e.
e. How far away is the lake?

4. Tennis stroke. A 7 cm diameter tennis ball is hit by a 28 cm diameter tennis racket. At t_0=0, when the racket is exerting the maximum force on the ball, the racket is moving horizontally, the strings of the racket are stretched the maximum amount they will be stretched, and the rubber of the ball is flattened the maximum amount it will be flattened.
a. Draw a sketch at a scale of 1:2 (1 cm on the sketch is 2 cm in real life) of the racket and the ball at t_0.
b. Describe in several sentences what is happening at the time of the sketch. You might include a description of the shapes of the ball and the racket, forces, velocities, accelerations, and so forth.
c. Describe any assumptions you are making about quantities whose values are not given in the problem.
d. Draw, on the sketch, symbols and vectors showing all of the forces on the ball and on the racket.
e. Draw, on the sketch, symbols and vectors showing the velocity and acceleration of the ball and of the racket.

5. Tomato soup. A can of Campbell's Tomato Soup is 67 mm in diameter and 98 mm high. After the soup is poured out of the can into a pot, a layer of soup 0.4 mm thick remains on all of the inside surfaces of the can. What fraction of the soup remains in the can?

6. Mushroom soup. The height of a can of mushroom soup is 1.6 times the diameter of the bottom of the can. When the soup is poured out of the can into a pot, a layer of soup that has a thickness of 1% of the diameter of the bottom remains on all of the inside surfaces of the can. What fraction of the soup went into the pot?

7. Weather balloon. A weather balloon made of rubber with a density of 1 100 kg/m^3 has a mass of 14 kg. When inflated with helium its diameter is 8 meters.
a. What is the surface area of the inflated balloon?
b. What is the volume of the rubber of the inflated balloon?
c. What is the thickness of the rubber of the inflated balloon? (The volume of the rubber is equal to the surface area times the thickness. Why?)

8. Phono plug. The outside part of a gold-plated phono plug is a thin-walled hollow cylinder 27 mm long and 8 mm in diameter. The inside and outside surfaces of the cylinder are plated with an 0.8 μm thick layer of solid gold. When gold costs $400 per troy ounce (31.1 grams), what is the value of the gold on the cylinder?

9. Satellite height. A satellite is in a circular orbit that passes over New York and Los Angeles. At one time in its orbit, it is on the horizon when seen from New York and is also on the horizon when seen from Los Angeles.

a. Write a heading, draw a diagram, and label it with all of the symbols for all of the quantities you will need in order to answer part d.
b. Make a list of definitions, symbols, values, and units for all of the quantities you will need to answer part d.
c. Make the preliminary unit conversion and geometric calculations that will help to solve part d.
d. What is the height of the satellite's orbit above the surface of the earth?

10. Filling stadium. A football stadium with 35 000 seats has three gates. When gate A alone is open it takes 40 minutes to fill the stadium. When gate B alone is open it takes 40 minutes to fill the stadium. When gate C alone is open it takes 60 minutes to fill the stadium. How long does it take to fill the stadium with all three gates open?

11. Skydiver. When a skydiver jumps out of an airplane, a surveyor on the ground records these values of her height as a function of time:

time	height
0 s	1400 m
1 s	1386 m
2 s	1365 m
3 s	1339 m
4 s	1308 m
5 s	1275 m
6 s	1240 m
7 s	1204 m
8 s	1167 m
9 s	1129 m
10 s	1091 m

time	height
15 s	899 m
20 s	706 m
25 s	512 m
30 s	318 m
35 s	282 m
40 s	244 m
45 s	205 m
50 s	167 m

a. Make a graph of her height (from 0 to 1 400 m) against time (from 0 to 80 seconds.)
b. Using the graph, describe her jump in several sentences. Tell about her position and her velocity. Describe what happened at 30 s. Estimate the time at which she will reach the ground.
c. Using changes in height divided by changes in time, make a table of her velocity vs. time. Use your graph of part b to help estimate values.
d. Make a graph of her velocity vs. time.

12. Same skydiver. The skydiver in the previous problem weighs 61 kg.
a. Discuss in one or more sentences the acceleration of the skydiver vs. time. Is there any time at which her acceleration is upward?
b. Discuss in one or more sentences the downward force, the upward force, and the total force on the skydiver between the time she leaves the airplane and the time she reaches the ground. (What is the downward force on the

skydiver? During those times when her acceleration is zero, what is the total force on her, and what is the upward force on her?)

13. Ping-pong balls on swimming pool. You want to keep your swimming pool warm by covering it with a single layer of ping-pong balls, packed as closely as possible. The pool is 5 m by 15 m and each ball is 2.5 cm in diameter.
a. Write a heading, draw a diagram, and label it with all of the symbols for all of the quantities you will need in order to answer part d.
b. Make a list of definitions, symbols, values, and units for all of the quantities you will need to answer part d.
c. Make the preliminary unit conversion and geometric calculations that will help to solve part d.
d. How many balls cover the pool and what fraction of the area of the pool is covered by the balls?

14. Building heat loss. A flat-roofed apartment building is 20 m wide, 23 m deep and 27 m high. On a winter day it loses heat through its surfaces exposed to the air at an average rate of 10 joules per square meter per second.
a. Write a heading, draw a diagram of the building, and label it with symbols for all of the quantities you will need in order to answer part d.
b. Make a list of definitions, symbols, values, and units for all of the quantities you will need to answer part d.
c. Write the preliminary geometric equations that will help to answer part d.
d. How much energy (in joules) does the building lose each day?

15. Memory refresh. A computer memory chip containing one million bits of random access memory must refresh every bit every 5 milliseconds. The chip is able to refresh one thousand bits in 50 nanoseconds. It is not available to the computer while it is refreshing itself. What fraction of the time is it available to the computer?

16. Fast car. A new Ferrari weighs 3 500 lb and can go from 0 to 60 miles per hour with constant acceleration in 4.2 seconds.
a. What is its acceleration in m/s^2?
b. What is the ratio of a to g, the earth's gravitational constant?
c. What is the forward force on the car while it is accelerating?
d. What is the ratio of the forward force on the car to its weight?
e. With the same acceleration, how long would it take the car to go one-quarter of a mile from a standing start?

17. Traveling piston. A piston in a 2 liter automobile engine travels 60 mm down and 60 mm back up each time the engine crankshaft makes one revolution. When the automobile is traveling at 55 miles per hour, the engine is turning at 2 600 revolutions per minute. What is the total distance the piston travels (relative to the cylinder wall) when the car travels 1 meter?

18. Automobile engine efficiency A car engine that can produce a maximum of 120 horsepower burns gasoline and turns the heat from the gasoline into useful work with an efficiency of 11 percent. The gasoline weighs 0.9 kg per liter and has a heat energy content of 45 megajoules per kilogram. How long can the engine run on one liter of gasoline while producing its maximum horsepower?

19. Pulsed laser. A laser turns electrical energy into light with an efficiency of 1%. A 1 000 microfarad capacitor charged to 5 000 volts transfers its stored electrical energy of 1.25×10^4 joules into the laser, which then produces a uniform 2×10^{-8} second long light pulse.
a. What is the energy in the resulting light pulse?
b. What is the light power during the pulse?

20. Tire wear. An automobile tire wears off 1/2 inch of tread thickness while traveling 50 000 miles. What is the average thickness (in meters) worn off in each revolution of the tire?

21. Passing cars. A Porsche traveling at a constant speed of 90 km/hr comes up behind a Corvette stopped at a light. When the Porsche is 30 m behind the Corvette, the light changes and the Corvette accelerates at 0.8 g. (g is the earth's gravitational constant.)
a. How far from the light does the Porsche pass the Corvette?
b. How far from the light does the Corvette pass the Porsche?
c. Plot on one graph the curve of position vs. time for each car.

22. Pole vault height. The height a pole vaulter can clear is mostly determined by the speed of his initial run. If all of his running kinetic energy is converted into potential energy during the jump, plot a graph of the height (in meters) he can clear vs. his horizontal velocity (in meters/second) when he leaves the ground. Use a reasonable range for his velocity.

23. Human-powered plane. A 220 lb human-powered airplane accelerates from rest along its runway with a constant acceleration of 0.5 m/s^2. The force on its wings is Kv^2 where v is the velocity of the plane and K=30 N-s^2/m^2. The plane just leaves the ground 33 meters from where it started. In what direction is the force on the wings?

24. Orbit radius and period. A satellite with an orbit radius of 10 000 km makes one complete trip around the earth in 9 940 seconds. How long does it take a satellite with an orbit radius of 20 000 km to make one complete trip? (It is not necessary to look up the mass of the earth or the gravitational constant to do this problem.)

25. Earthly speed. The earth revolves on its axis once each day while it orbits the sun once each year. Seen from the north pole of the earth, both of these motions are counterclockwise.

a. How fast does the center of the earth move in its orbit about the sun?

b. With respect to the center of the earth how fast does a person standing on the equator move because of the earth's rotation?

c. How fast does a person at your latitude move?

d. Assuming that the earth's axis is perpendicular to its orbit plane, what are your maximum and minimum speeds with respect to the sun?

e. At what time of day do you have your maximum speed?

Further Reading

Some books that are close in level and emphasis to *How to Solve Problems* are:

Becoming a Master Student, 10th Edition, David B. Ellis, Doug Toff, Dave Ellis, Houghton Mifflin, 2000, ISBN:0618206787. Although many believe that being a student, writing English, and driving a car are skills we all are born with, they are not. Get and read it, and learn how to be a student.

Used Math; For the First Two Years of College Science, 2nd Edition, Clifford E. Swartz, American Association of Physics Teachers, One Physics Ellipse, College Park, MD 20740-3845, 1993, ISBN:0917853504, A useful handbook of introductory applied mathematics containing the tools one needs to solve beginning science and engineering problems with confidence.

Succeed With Math; Every Student's Guide to Conquering Math Anxiety. Sheila Tobias, MacMillan Publishing Co. 1988, ISBN:0874472636. Another useful handbook of introductory applied mathematics written with sensitivity and style. Worth owning.

The definition of problem solving is expanded to the creative solution of mathematical problems by:

The Art of Problem Posing, 2nd Edition, Stephen I. Brown and Marion I. Walter, Lawrence Erlbaum Associates, 365 Broadway, Hillsdale, NJ 07642, 1991, ISBN:0805802584. Describes how good mathematicians learn to see problems where no one knows that there are problems, by deriving general rules from particular examples. Shows that a problem is well solved if the solution leads to an even more important unsolved problem. Essential ideas for those who plan to become mathematicians.

How to Solve It; A New Aspect of Mathematical Method, George Polya and Gyorgy Polya. Princeton University Press, 1988, ISBN:0691023565. The granddaddy of them all. This discussion of mostly geometric and algebraic problem solving has led to most of the other books on problem solving. A how-to-do-it book that contains every method one can think of.

Mathematical Discovery: On Understanding, Learning, and Teaching Problem Solving, George Polya, John Wiley and Sons, 1981. Greatly expanded how-to-solve-it book aimed at teachers of mathematics and of problem solving. Many mathematical examples.

Mathematical Problem Solving, Alan H. Schoenfeld, Academic Press, 1985, ASIN:0126288712. Scholarly advanced treatise with a scholarly advanced style. Good blow-by-blow descriptions of students' attempts to solve geometrical problems with an analysis of the thought processes of different types of problem solvers.

Problem Solving and Comprehension, 6th Ed. Arthur Whimbey and Jack Lochhead, Lawrence Erlbaum Associates, 365 Broadway, Hillsdale, NJ 07642, 1999, ISBN:0805832742. Shows methods to use on math and logic problems of the type found on the SAT's, GRE's, and other tests. Can only help.

More general advice for success in research and engineering can be found in:

The Art of Scientific Investigation, 3rd edition, W. I. B. Beveridge, Random House, 1960, ASIN:0394701291. Beautifully written introduction to scientific thinking, containing many descriptions of how well-known scientists worked toward their discoveries. The best place to learn the real rewards of doing science.

Being Successful as an Engineer, W. H. Roadstrum, Engineering Press, P. O. Box 5, San Jose, CA 95103, 1978, ISBN:0910554242. A good introduction for people beginning an engineering career. Covers everything from proposals through research, manufacturing, and quality control with emphasis on the skillful management of an engineering group.

An Introduction to Scientific Research, E. Bright Wilson, Jr., McGraw Hill, 1952, Dover 1991, ISBN:0486665453. A wonderful handbook for Ph.D. students in the sciences. Goes from the choice of a research problem. through the design of experiments, statistics, data reduction, and mathematical calculations, to the reporting of scientific results.

James Adams's books are an introduction to the literature on creative thinking in engineering, science, art, music, and everyday life:

Conceptual Blockbusting: A Guide to Better Ideas, 4th Ed., James L. Adams, Perseus Publishing, 2001, ISBN:0738205370. A book, written by an engineer, that is a best-seller among managers, Adams encourages solving problems by taking creative and imaginative end runs around them. Look at this book to see that "solving problems" has many meanings, only one of which is discussed in *How to Solve Problems.*

The Care and Feeding of Ideas; A Guide to Encouraging Creativity, James L. Adams, Addison Wesley, 1987, ASIN:0201101602. Very broad non-technical discussion of how to solve problems that have not yet been formulated. Emphasizes freeing the mind from ruts and nurturing partially formed thoughts.

After you have solved a problem and want to describe it to others in clear English, read:

The Elements of Style, 4th Ed., William Strunk, Jr., and E. B. White, Allyn & Bacon, 2000, ISBN:020530902X. Buy it. Read it. Use it.

Style: Toward Clarity and Grace, Joseph M. Williams, University of Chicago Press, 1995, ISBN:0226899152. If the third edition of *How to Solve Problems* was better than the second, it is because of Williams's book.

Dazzle 'em With Style: The Art of Oral Scientific Presentation, Robert R. H. Anholt, W. H. Freeman and Co. 1994, ISBN:0716725835. A useful and informal guide to presenting talks for courses, thesis defenses, proposal reviews, and scientific meetings.

Problem Solutions

1. Satellite Period

(Your name and date)

A. DRAWING

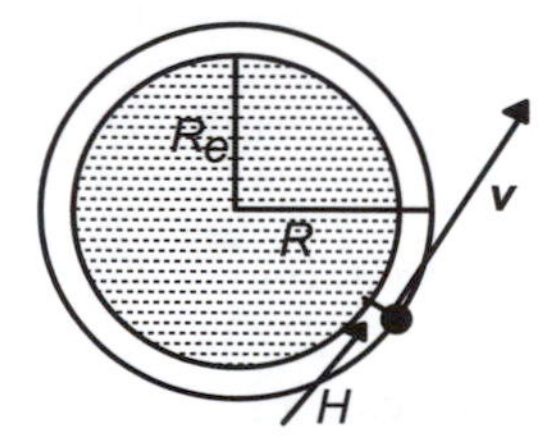

B. DEFINITIONS

Height of satellite above earth's surface	$= H$	=640	km
Velocity of satellite	$= v$	=7.54	km/s
Radius of earth	$= R_e$		m
Radius of satellite orbit	$= R$		m
Period of satellite orbit	$= P$		s

DATA

R_e = 6371 km — Handbook of Chemistry and Physics, 57th Ed. Chemical Rubber Co. 1971. pg F148

C. GEOMETRY

Find the orbit radius

$$R = R_e + H = 6371 \text{ km} + 640 \text{ km} = 7011 \text{ km}$$

Find the period

$$P = \frac{2\pi R}{v} = \frac{2\,\pi\; 7011 \text{ km}}{7.54 \text{ km/s}}$$

$$= 5842 \text{ s} = 5842 \text{ s}\left(\frac{1 \text{ min}}{60 \text{ s}}\right)$$

$$= 97 \text{ min} = 97 \text{ min}\left(\frac{1 \text{ hr}}{60 \text{ min}}\right) = 1.62 \text{ hr}$$

D. The period of the satellite is P = 5842 s or 97 min or 1.62 hr.

3. Distant Lake

(Your name and date)

A. DRAWING

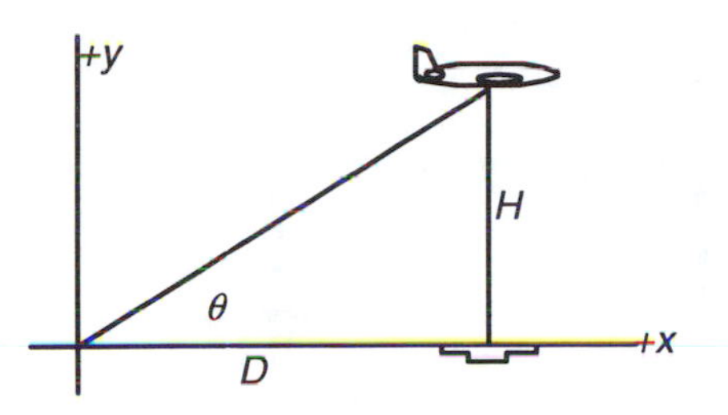

B. KNOWN QUANTITIES

Height of airliner	$= H$	= 10 000	m
Angle between horizontal and airliner	$= \theta$	= 32	°

C. UNKNOWN QUANTITIES

Distance to lake	$= D$		m

D. GEOMETRY

$$H/D = \tan\theta$$

$$D = \frac{H}{\tan\theta}$$

$$= \frac{10\,000 \text{ m}}{\tan 32°}$$

$$= \frac{10\,000 \text{ m}}{0.625} = 1.60 \times 10^4 \text{ m}$$

E. The distance to the lake is $D = 1.60 \times 10^4$ m or 16 km.

5. Tomato soup

(Name and date)

DRAWING

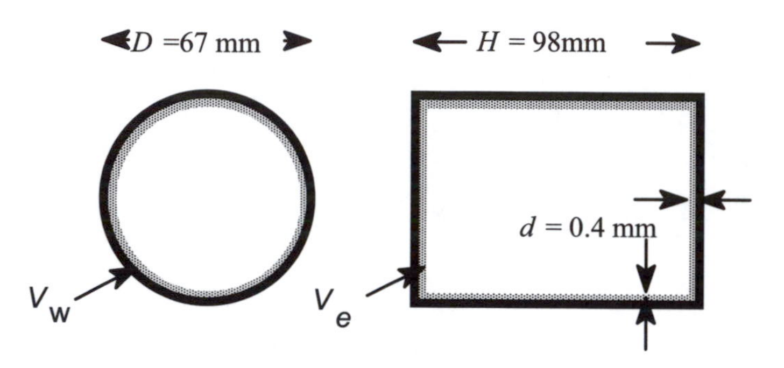

DEFINITIONS

Height of can	$= H$	$= 98$	mm
Diameter of can	$= D$	$= 67$	mm
Thickness of soup layer	$= d$	$= 0.4$	mm
Area of one end	$= A_e$		mm^2
Area of can wall	$= A_w$		mm^2
Volume of soup on one end	$= V_e$		m^3
Volume of soup on wall	$= V_w$		m^3
Volume of soup remaining in can	$= V$		m^3
Original volume of soup in can	$= V_o$		m^3
Ratio of remaining volume to original volume	$= R$		

UNIT CONVERSION

$$\mathrm{H} = 98 \text{ mm}\left(\frac{1 \text{ m}}{1000 \text{ mm}}\right) = 0.098 \text{ m}$$

$$D = 67 \text{ mm}\left(\frac{1 \text{ m}}{1000 \text{ mm}}\right) = 0.067 \text{ m}$$

$$d = 0.4 \text{ mm}\left(\frac{1 \text{ m}}{1000 \text{ mm}}\right) = 0.0004 \text{ m}$$

GEOMETRY

Find the area of one end of the can

$$A_e = \pi r^2 \qquad = \pi\left(\frac{D}{2}\right)^2 \qquad = \pi\frac{D^2}{4} \qquad = \frac{\pi}{4}D^2$$

$$= \frac{\pi}{4}(0.067\text{ m})^2 \qquad = (0.7854)(0.00449)$$

$$= 3.526\text{x}10^{-3}\text{ m}^2$$

Find the volume of soup on one end of the can

$$V_e = A_e\, d$$

$$= 3.526\text{x}10^{-3}\text{ m}^3\ (0.0004\text{ m})$$

$$= 1.41\text{ x }10^{-6}\text{ m}^3$$

Find the area of the wall of the can

$$A_w = 2\pi r H \qquad = 2\pi\frac{D}{2}H \qquad = \pi D H$$

$$= \pi(0.067\text{ m})(0.098\text{ m})$$

$$= 2.063\text{x}10^{-2}\text{ m}^2$$

Find the volume of soup on the wall

$$V_w = A_w d$$

$$= 2.063\text{x}10^{-2}\text{ m}^2\ (0.0004\text{ m})$$

$$= 8.251\text{x}10^{-6}\text{ m}^3$$

Find the total volume of the soup left in the can

$$V = Vw + 2Ve$$

$$= 8.251\text{x}10^{-6}\text{ m}^3 + 2(1.41\text{ x }10^{-6}\text{ m}^3)$$

$$= 1.107\text{x}10^{-5}\text{ m}^3$$

Find the original volume of soup

$$V_o = A_e H$$

$$= 3.526\text{x}10^{-3}\text{ m}^2\ (0.098\text{ m})$$

$$= 3.455\text{x}10^{-4}\text{ m}^3.$$

Find the ratio of the remaining soup to the original soup

$$R = \frac{V}{V_o}$$

$$= \frac{1.107\text{x}10^{-5}}{3.455\text{x}10^{-4}} \qquad = 3.204\text{x}10^{-2} \qquad = 3.2\%$$

The ratio of the volume of soup left in the can to the volume of the original soup is 0.032 or 3.2%.

7. Weather balloon

(Name and date)

DRAWING

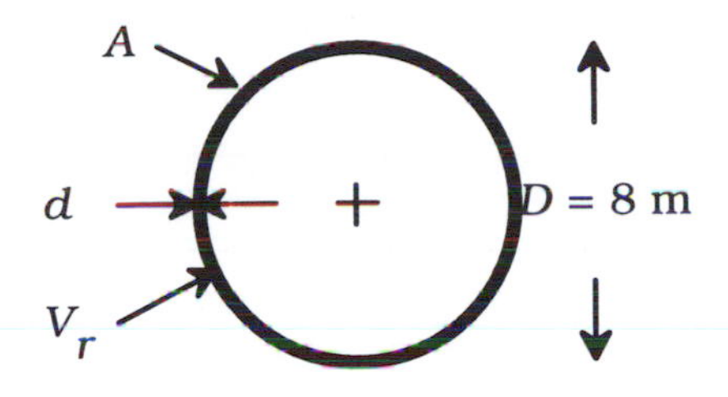

DEFINITIONS

Diameter of balloon	$= D$	$= 8$	m
Mass of balloon	$= m$	$= 14$	kg
Density of rubber	$= \rho$	$= 1100$	kg/m^3
Surface area of balloon	$= A$		m^2
Volume of rubber	$= V$		m^3
Thickness of rubber	$= d$		m

A. Find the surface area of the balloon

$$A = 4\pi r^2 = 4\pi\left(\frac{D}{2}\right)^2 = 4\pi\frac{D^2}{4} = \pi D^2$$

$$= \pi(8\text{ m})^2 = 2.011\text{x}10^2\text{ m}^2$$

The surface area of the balloon is A = 200 m^2

B. Calculate the volume of the rubber

$$V = Ad$$

Calculate the mass of the rubber

$$m = \rho V$$

Solve for the volume of the rubber

$$V = \frac{m}{\rho} = \frac{14\text{ kg}}{1100\text{ kg/m}^3} = 1.273\text{x}10^{-2}\text{ m}^3$$

The volume of the rubber is V = 1.3 x 10^{-2} m^3.

C. Solve for the thickness of the rubber

$$d = \frac{V}{A} = \frac{1.273 \times 10^{-2}\ \text{m}^3}{2.011 \times 10^{2}\ \text{m}^2} = 6.33 \times 10^{-5}\ \text{m}$$

Check by using the equations for V and A

$$d = \frac{V}{A} = \frac{m/\rho}{A} = \frac{m/\rho}{\pi D^2}$$

$$= \frac{m}{\pi \rho D^2}$$

$$= \frac{14}{\pi\ 1100\ (8)^2} = 6.33 \times 10^{-5}\ \text{m}$$

The rubber of the balloon is 6.3×10^{-5} m or 63 μm thick.

9. Satellite Height.

(Name and date)

A.

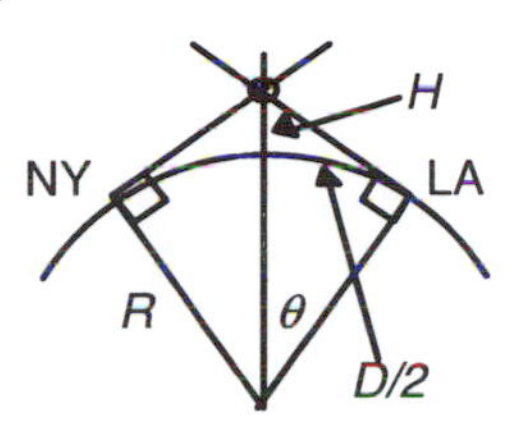

B. DEFINITIONS

Distance from LA to NY	$= D$		mi
Radius of earth	$= R$	$= 6.37\text{x}10^6$	m
Angle at center of earth between vector to satellite and vector to LA	$= \theta$		radians
Height of satellite	$= H$		m

DATA

$D \quad = 2794$ mi — from Rand McNally Road Atlas of US, Canada, and Mexico. 1983, pg. 105

UNITS

$$D = 2794 \text{ mi} = 2794 \text{ mi}\left(\frac{1609.3 \text{ m}}{1 \text{ mi}}\right) = 4.496\text{x}10^6 \text{ m}$$

C. Find angle at center of earth

$$\theta = \frac{D/2}{R} = \frac{4.496\text{x}10^6}{2(6.37\text{x}10^6)} = 0.3529 \text{ radians}$$

D. Find height of satellite using right triangle

$$\cos\theta = \frac{R}{R+H} = \frac{1}{1+H/R}$$

$$(1+H/R)\cos\theta = 1$$

$$1 + H/R = \frac{1}{\cos\theta}$$

$$\frac{H}{R} = \frac{1}{\cos\theta} - 1$$

$$= \frac{1}{\cos 0.3529 \text{ rad}} - 1 = \frac{1}{0.9384} - 1$$

$$= 1.06567 - 1 = 0.06567$$

$$H = 0.06567\, R = 0.06567(6.37\text{x}10^6 \text{ m})$$

$$= 4.18\text{x}10^5 \text{ m} = 418 \text{ km}$$

The height of the satellite is $H = 418$ km.

11. Sky Diver

(Name and Date)

DRAWING

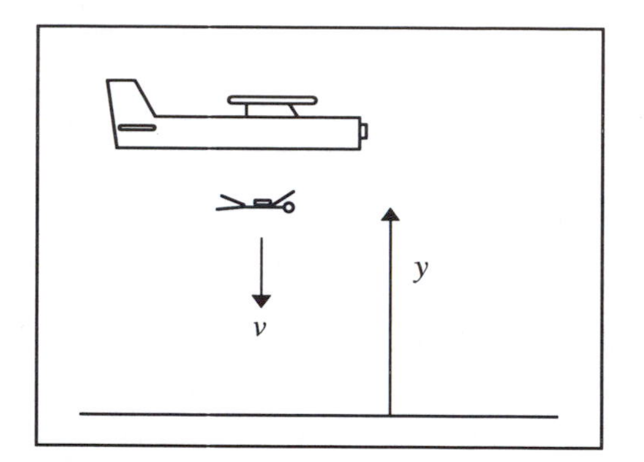

DEFINITIONS

Time after diver leaves plane	$= t$	s
Distance between diver and ground	$= y$	m
Velocity of diver	$= v$	m/s

C. DATA

ti [s]	heigh [m]	tim [s]	vel [m/s
0	14	0.	-
1	13	1.	-
2	13	2.	-
3	13	3.	-
4	13	4.	-
5	12	5.	-
6	12	6.	-
7	12	7.	-
8	11	8.	-
9	11	9.	-
1	10	1	-

tim [s]	heigh [m]	time [s]	vel [m/s
1	899	17.	-
2	706	22.	-
2	512	27.	-
3	318	32.	-7
3	282	37.	-8
4	244	42.	-8
4	205	47.	-8
5	166	52.	-8
5	127	57.	-8
6	88	62.	-8
6	49	67.	-8
7	10	71.	-8
7	0	74.	0
7	0	77.	0
8	0		

$v = \Delta y/\Delta t$. The velocity at 0.5 second was found by subtracting the height at 1 s from the height at 0 s and dividing by the time difference of 1 s.

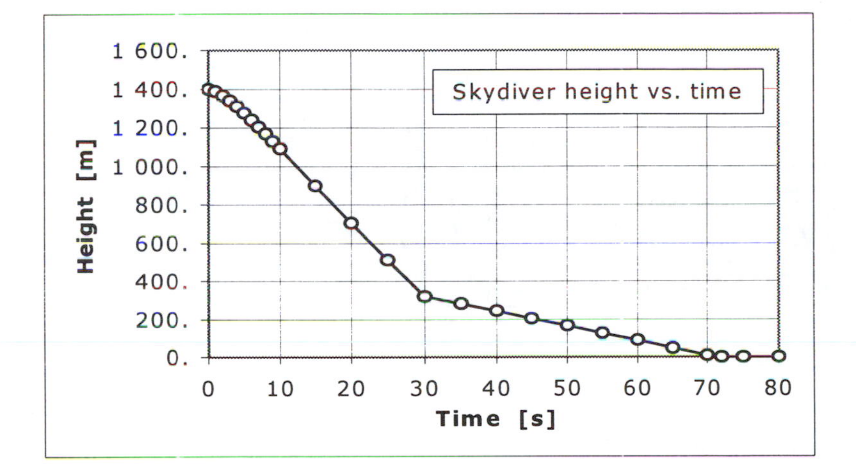

B. At t=0, the diver leaves the plane with zero vertical velocity and accelerates downward for about 5 seconds. She then falls at constant velocity until she opens her parachute at 30 seconds, when she is about 300 meters above the ground. She quickly comes to a new, lower constant velocity and floats down until she hits the ground at about 72 seconds.

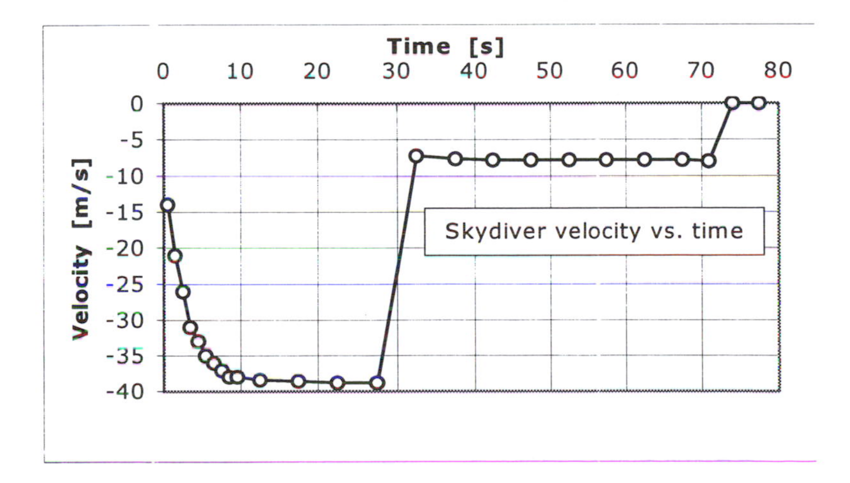

D. The curve of velocity vs. time shows that her velocity increases until about 9 sec, at which time she is travelling downward at about 38 m/s. After the parachute opens, she falls at about 7 m/s. The velocity graph is not reliable near 30 seconds and again near 72 seconds because the data points are too far apart in time to show the actual motion.

13. Ping pong balls on swimming pool

(Name and date)

A. DRAWING

B. DEFINITIONS

Width of pool	$= W$	$= 5$	m
Length of pool	$= L$	$= 15$	m
Diameter of one ball	$= D$	$= 2.5$	cm
Radius of one ball	$= R$		m
Cross-section area of one ball	$= A_1$		m^2
Spacing of ball centers	$= S$		m
Number of balls per meter of width	$= N_W$		
Number of balls per meter of length	$= N_L$		
Total number of balls	$= N$		
Total cross-sectional area of all balls	$= A_b$		m^2
Surface area of pool	$= A_p$		m^2
Fraction of surface covered by balls	$= F$		

C. UNIT CONVERSION

$$D = 2.5 \text{ cm} = 2.5 \text{ cm}\left(\frac{1 \text{ m}}{100 \text{ cm}}\right) = 0.025 \text{ m}$$

D. Find fraction of pool surface covered by balls.

Find ball radius

$$R = D/2 = 0.025 \text{ m}/2 = 0.0125 \text{ m}$$

Find spacing between rows of balls

$$S = \sqrt{(2R)^2 - R^2} = \sqrt{4R^2 - R^2} = \sqrt{3R^2}$$

$= \sqrt{3}\ R \qquad = 1.732\, R \qquad = 1.732\ (0.0125)$

$= 0.02165$ m

Find number of balls along length of pool

$$N_L = \frac{L}{D} = \frac{15\text{ m}}{0.025\text{ m}} = 600 \text{ balls}$$

Find number of rows along width of pool

$$N_W = \frac{W}{S} = \frac{5\text{ m}}{0.02165} = 230.95 = 230 \text{ balls}$$

Find total number of balls

$$N = N_L N_W = 600\ (230) = 138\ 000 \text{ balls}$$

D. The total number of balls is N = 138 000.

Find cross section area of one ball

$$A_1 = \pi R^2 = \pi\ (0.0125)^2 = 4.91\text{x}10^{-4}\text{ m}^2$$

Find total cross sectional area of all balls

$$A_b = NA_1 = 138\ 000\ (4.91\text{x}10^{-4}\text{ m}^2) = 67.76\text{ m}^2$$

Find total surface area of pool

$$A_p = LW = 15\text{ m}\ (5\text{ m}) = 75\text{ m}^2$$

Find fraction of pool surface covered by balls

$$F = \frac{A_b}{A_p} = \frac{67.76\text{ m}^2}{75\text{ m}^2} = 0.90$$

Check algebraically

$$F = \frac{A_b}{A_p} = \frac{A_1 N}{LW} = \frac{\pi R^2 N_L N_W}{LW}$$

$$= \frac{\pi R^2 (L/D)(W/S)}{LW} = \frac{\pi R^2}{SD}$$

$$= \frac{\pi R^2}{\sqrt{3}R\ 2R} = \frac{\pi}{2\sqrt{3}} = 0.907$$

The fraction of the surface of the pool that is covered by balls is F = 0.91.

Since the Rs cancelled in the last equation, the fraction covered doesn't depend on the size of the balls or the size of the pool (as long as the balls are much smaller than the pool.) Can you find a way to solve this problem without using the size of the pool at all?

15. Memory refresh

(Name and date)

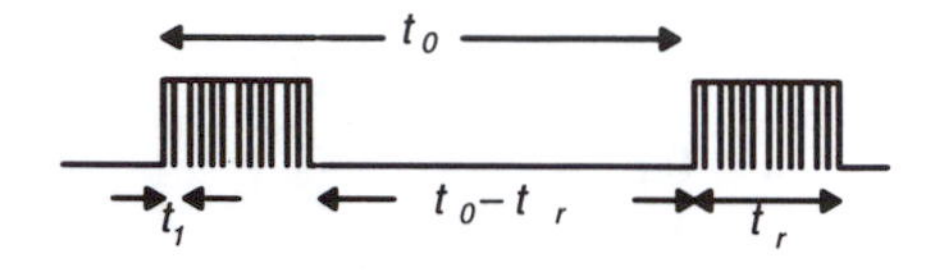

DEFINITIONS

Total number of bits	$= N$	= 1 000 000	
Time between refreshes	$= t_0$	= 5	ms
Time to refresh one batch	$= t_1$	= 50	ns
Number of bits in one batch	$= N_1$	= 1 000	
Number of batches	$= N_b$		
Total refresh time	$= t_r$		s
Total useful (non-refresh) time	$= t_u$		s
Fraction of useful time	$= F$		

UNIT CONVERSION

$$t_0 \quad = 5 \text{ ms} \qquad = 5 \text{ ms} \left(\frac{10^{-3} \text{ s}}{1 \text{ ms}}\right) \qquad = 5\text{x}10^{-3} \text{ s}$$

$$t_1 \quad = 50 \text{ ns} \qquad = 50 \text{ ns} \left(\frac{10^{-9} \text{ s}}{1 \text{ ns}}\right) \qquad = 5\text{x}10^{-8} \text{ s}$$

Find number of batches

$$N_b \quad = N/N_1 \qquad = 1\,000\,000/1000 \qquad = 1000$$

Find total refresh time

$$t_r \quad = N_b\, t_1 \qquad = 1\,000\,(5\text{x}10^{-8}\text{s}) \qquad = 5\text{x}10^{-5} \text{ s}$$

Find total useful time

$$t_u \quad = t_0 - t_r \qquad = 5\text{x}10^{-3} \text{ s} - 5\text{x}10^{-5} \text{ s}$$

$$= 4.95\text{x}10^{-3} \text{ s}$$

Find fraction of time computer is available

$$F \quad = \frac{t_u}{t_0} \qquad = \frac{4.95\text{x}10^{-3}}{5\text{x}10^{-3}} = 0.99$$

The fraction of time that the memory is available is F =0.99.

17. Travelling Piston

(Name and date)

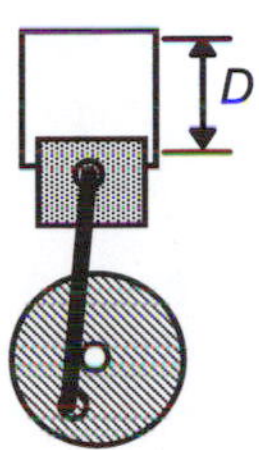

DEFINITIONS

Distance piston travels in 1/2 revlution	$= D$	$= 60$	mm
Speed of car	$= v$	$= 55$	mi/hr
Engine revolutions per minute at 55 mph	$= f$	$= 2600$	min^{-1}
Distance car moves	$= x_c$	$= 1$	m
Distance piston moves	$= x_p$		m
Number of engine revolutions for 1 m of car movement	$= N$		

UNIT CONVERSION

$$D \quad = 60\text{ mm} \qquad = 60\text{ mm}\left(\frac{1\text{ m}}{1000\text{ mm}}\right) \qquad = 0.06\text{ m}$$

$$f \quad = 2600\,\frac{\text{rev}}{\text{min}} \qquad = 2600\,\frac{\text{rev}}{\text{min}}\left(\frac{1\text{ min}}{60\text{ s}}\right) \qquad = 43.33\text{ rev/s}$$

$$v \quad = 55\,\frac{\text{mi}}{\text{hr}} \qquad = 55\,\frac{\text{mi}}{\text{hr}}\left(\frac{1609.3\text{ m}}{1\text{ mi}}\right)\left(\frac{1\text{ hr}}{3600\text{ s}}\right)$$

$$= 24.59\text{ m/s}$$

Find the number of engine revolutions for 1 m of car movement (in 1 s the engine makes f revolutions and the car moves v meters)

$$N \quad = \frac{f}{v} \qquad = \frac{43.33\text{ rev/s}}{24.59\text{ m/s}} \qquad = 1.762\text{ rev/m}$$

Find the distance the piston moves when the car moves 1 m

$$x_p \quad = 2DNx_c \qquad = 2\,(0.06\text{ m})(1.762\text{ m}^{-1})(1\text{ m})$$

$$= 0.21\text{ m}$$

When the car moves 1 m, the piston moves 0.21 m.

19. Pulsed Laser

(Name and date)

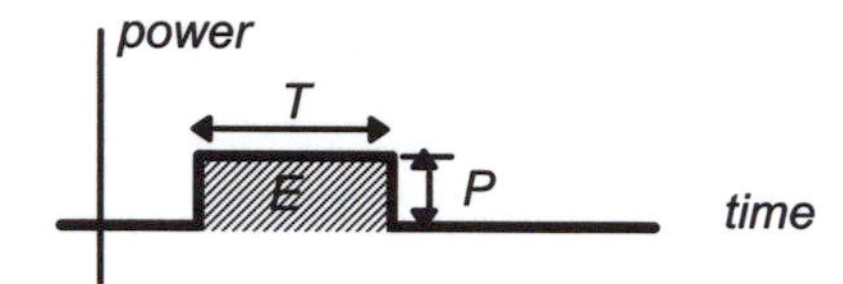

DEFINITIONS

Length of pulse	$= T$	$= 2\text{x}10^{-8}$	s
Electrical energy in pulse	$= E_e$	$= 1.25\text{x}10^4$	J
Light energy in pulse	$= E_l$		J
Efficiency	$= e$	$= 1\%$	
Power during pulse	$= P$		W

unit conversion

$$e = 1\% = 1\%\left(\frac{1}{100\%}\right) = 0.01$$

A. Find light energy in pulse

$$E_l = eE_e = 0.01\ (1.25 \text{ x } 10^4)$$

$$= 1.25 \text{ x } 10^2 \text{ J}$$

The light energy in one pulse is $E_l = 126$ J.

B. Find power during pulse

$$P = \frac{E_l}{T} = \frac{eE_e}{T} = \frac{0.01\ (1.25\text{x}10^4)}{2\text{x}10^{-8}}$$

$$= 6.25\text{x}10^9 \text{ W}$$

Laser power during the pulse is $P = 6.25\text{x}10^9$ W.

21. Passing cars

(Name and date)

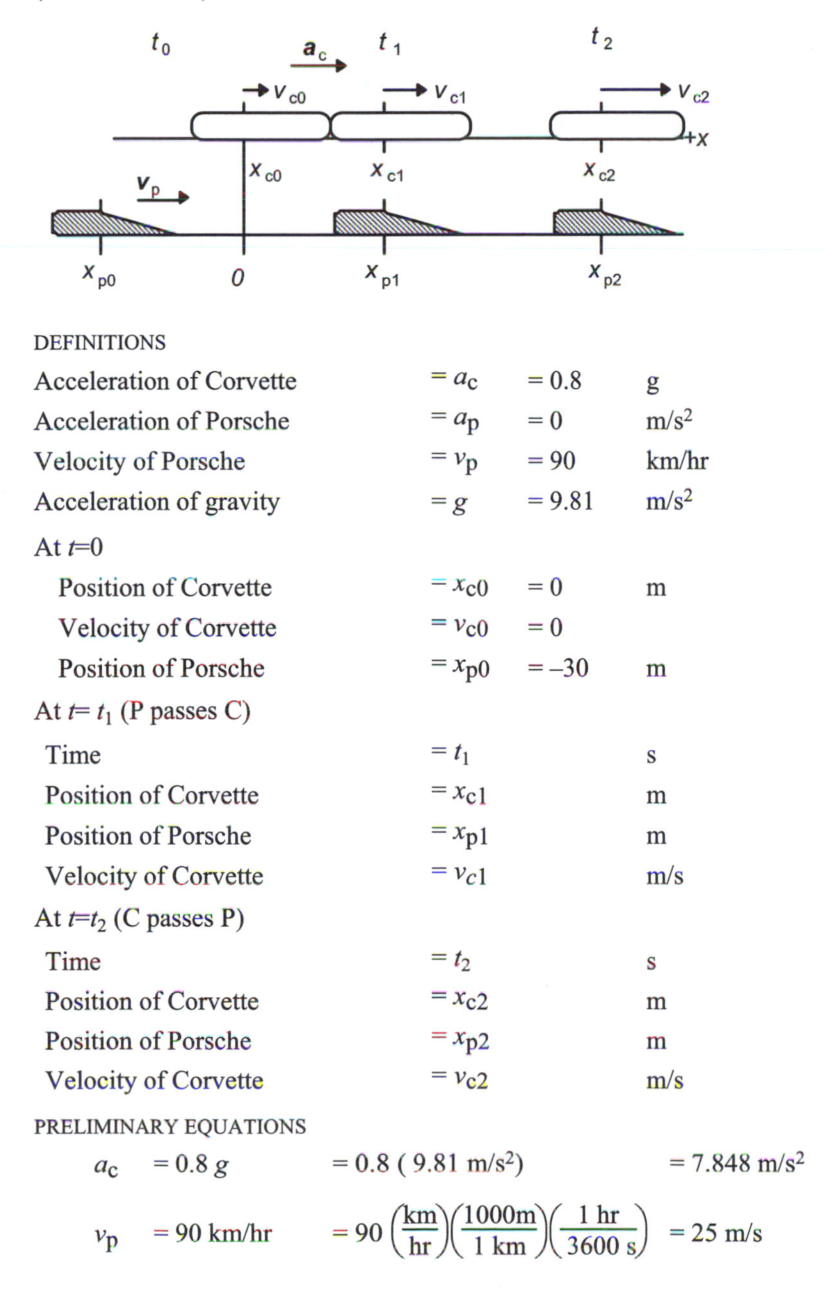

DEFINITIONS

Acceleration of Corvette	$= a_c$	$= 0.8$	g
Acceleration of Porsche	$= a_p$	$= 0$	m/s^2
Velocity of Porsche	$= v_p$	$= 90$	km/hr
Acceleration of gravity	$= g$	$= 9.81$	m/s^2
At t=0			
Position of Corvette	$= x_{c0}$	$= 0$	m
Velocity of Corvette	$= v_{c0}$	$= 0$	
Position of Porsche	$= x_{p0}$	$= -30$	m
At $t = t_1$ (P passes C)			
Time	$= t_1$		s
Position of Corvette	$= x_{c1}$		m
Position of Porsche	$= x_{p1}$		m
Velocity of Corvette	$= v_{c1}$		m/s
At t=t_2 (C passes P)			
Time	$= t_2$		s
Position of Corvette	$= x_{c2}$		m
Position of Porsche	$= x_{p2}$		m
Velocity of Corvette	$= v_{c2}$		m/s

PRELIMINARY EQUATIONS

$$a_c = 0.8\,g = 0.8\,(9.81\text{ m/s}^2) = 7.848\text{ m/s}^2$$

$$v_p = 90\text{ km/hr} = 90\left(\frac{\text{km}}{\text{hr}}\right)\left(\frac{1000\text{m}}{1\text{ km}}\right)\left(\frac{1\text{ hr}}{3600\text{ s}}\right) = 25\text{ m/s}$$

$$v = v_0 + at$$

$$x = x_0 + v_0 t + \tfrac{1}{2} at^2$$

When the Porsche passes the Corvette, $t = t_1$, and $x_{p1} = x_{c1}$

For the Corvette

$$x_{c1} = x_{c0} + v_{c0}t_1 + \tfrac{1}{2} a_c t_1^2 \qquad = 0 + 0 + \tfrac{1}{2} a_c t_1^2$$

For the Porsche

$$x_{p1} = x_{p0} + v_p t_1 + \tfrac{1}{2} a_p t_1^2 \qquad = x_{p0} + v_p t_1 + 0$$

Equate x_{c1} and x_{p1}

$$x_{c1} = x_{p1}$$

$$\tfrac{1}{2} a_c t_1^2 = x_{p0} + v_p t_1$$

Solve for t_1

$$\tfrac{1}{2} a_c t_1^2 - v_p t_1 - x_{p0} = 0$$

This looks like

$$at^2 + bt + c = 0$$

So

$$t = \frac{-b \pm \sqrt{b^2 - 4ac}}{2a}$$

And

$$t_1 = \frac{+v_p \pm \sqrt{v_p^2 - 4\,(\tfrac{1}{2}a_c)(-x_{p0})}}{a_c}$$

$$= \frac{v_p \pm \sqrt{v_p^2 + 2a_c x_{p0}}}{a_c}$$

$$= \frac{25\text{ m/s} \pm \sqrt{(25\text{ m/s})^2 + 2\,(7.848\text{ m/s}^2)\,(-30\text{ m})}}{7.848\text{ m/s}^2}$$

$$= \frac{25 \pm \sqrt{625 - 470.9}}{7.848} = \frac{25 \pm \sqrt{154.1}}{7.848} \qquad = \frac{25 \pm 12.4}{7.848}$$

To find t_1, the smaller root

$$t_1 \quad = \frac{25 - 12.4}{7.848} \qquad = \frac{12.6}{7.848} \qquad = 1.60 \text{ s}$$

To find t_2, the larger root

$$t_2 \quad = \frac{25 + 12.4}{7.848} \qquad = \frac{37.4}{7.848} \qquad = 4.77 \text{ s}$$

A. To find x_1, the position at which the Porsche passes the Corvette

$$x_{c1} \quad = \tfrac{1}{2} a_c t_1^2 \qquad = \tfrac{1}{2}(7.848 \text{ m/s}^2)\,(1.60 \text{ s})^2 \qquad = 10.05 \text{ m}$$

Check using position of Porsche at t_1

$$x_{p1} \quad = x_{p0} + v_p t_1 \qquad = --30 \text{ m} + 25 \text{ m/s}\,(1.60 \text{ s}) \qquad = 10.0 \text{ m}$$

B. To find x_2, the position at which the 'vette passes the Porsche

$$x_{c2} \quad = \tfrac{1}{2} a_c t_2^2 \qquad = \tfrac{1}{2}(7.848 \text{ m/s}^2)\,(4.77 \text{ s})^2 \qquad = 89.28 \text{ m}$$

Check using position of Porsche at t_2

$$x_{p2} \quad = x_{p0} + v_p t_2 \qquad = -30 \text{ m} + 25 \text{ m/s}\,(4.77 \text{ s}) \qquad = 89.25 \text{ m}$$

The position at which the Porsche passes the Corvette is $x_1 = 10.0$ m.

The position at which the Corvette passes the Porsche is $x_2 = 89.3$ m.

C.

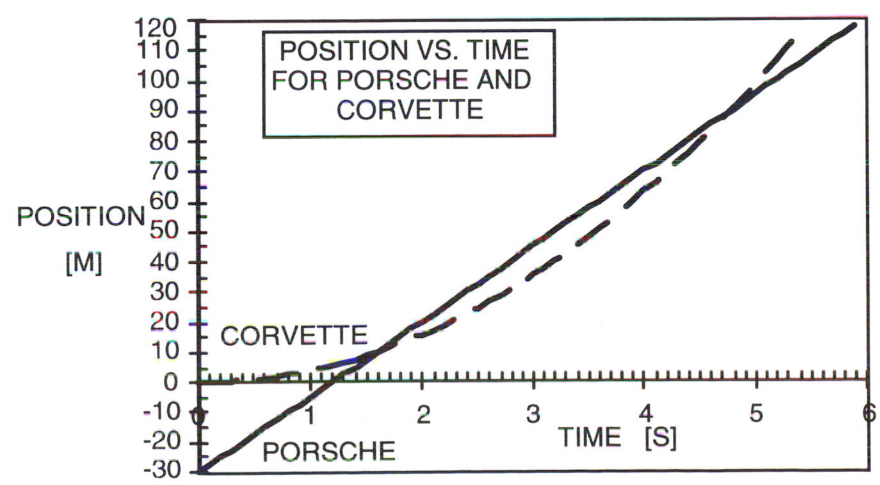

What is the time at which the Corvette gets a speeding ticket?

23. Human-powered plane.

(Name and date)

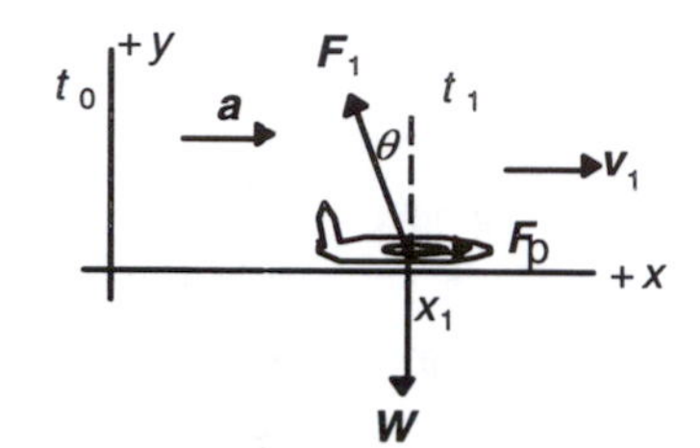

DEFINITIONS

Weight of plane	$= W$	$= 220$	lb
Horizontal acceleration of plane	$= a_x$	$= 0.5$	m/s
Vertical acceleration of plane	$= a_y$	$= 0$	m/s^2
Constant for force on wings	$= K$	$= 30$	N-s^2/m^2
At t = 0			
Horizontal position of plane	$= x_0$	$= 0$	m
Velocity of plane	$= v_0$	$= 0$	m/s
At t_1 (when plane just lifts off ground)			
Time	$= t_1$		sec
Position	$= x_1$	$= 33$	m
Velocity	$= v_1$		m/s
Force on wings	$= F_1$		N
Upward component of force	$= F_y$		N
Angle of force from vertical	$= \theta$		°

PRELIMINARY EQUATIONS

$$W = 220 \text{ lb} \qquad = 220 \text{ lb}\left(\frac{4.448 \text{ N}}{1 \text{ lb}}\right) \qquad = 978.6 \text{ N}$$

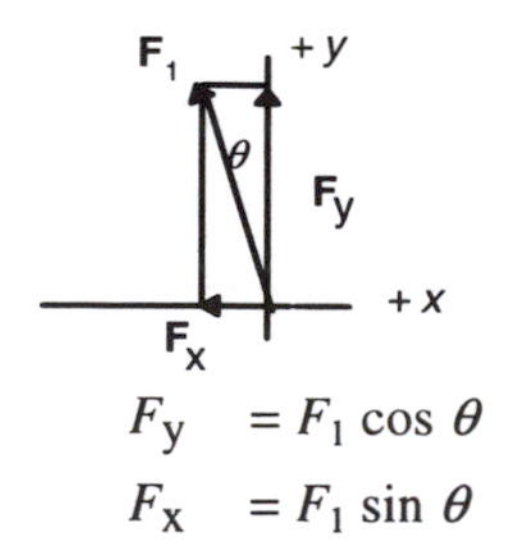

$$F_y = F_1 \cos\theta$$

$$F_x = F_1 \sin\theta$$

At takeoff time t_1

$$F_y = W$$

$$F_1 = Kv_1^2$$

$$\frac{F_y}{\cos\theta} = Kv_1^2$$

$$\cos\theta = \frac{F_y}{Kv_1^2} \qquad = \frac{W}{Kv_1^2}$$

To find v_1

For motion with constant acceleration

$$a = \text{constant}$$

$$v = v_0 + at$$

$$x = x_0 + v_0 t + \frac{1}{2}at^2$$

At t_1, when the plane leaves the ground

$$x_1 = 0 + 0 + \frac{1}{2}a_x t_1^2$$

$$t_1^2 = \frac{2x_1}{a_x}$$

Find v_1 in terms of t_1

$$v_1 = 0 + a_x t_1$$

But we need v_1^2

$$v_1^2 = a_x^2 t_1^2 \qquad = a_x^2\left(\frac{2x_1}{a_x}\right) \qquad = 2a_x x_1$$

Back to $\cos\theta$

$$\cos\theta = \frac{W}{Kv_1^2} \qquad = \frac{W}{K\,(2a_x x_1)}$$

$$= \frac{978.6\ \text{N}}{(30\ \text{N-s}^2/\text{m}^2)\ 2\ (0.5\ \text{m/s}^2)\ (33\ \text{m})} = 0.9885$$

$$\theta = 8.7\,°$$

The angle of the force from the vertical is $\theta = 8.7°$.

25. Earthly Speed

(Name and date)

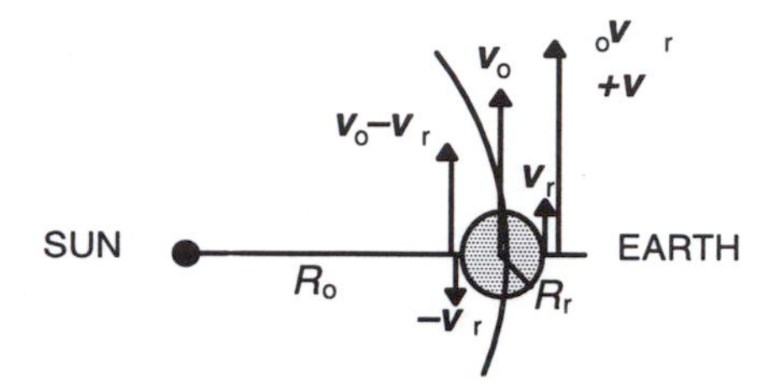

DEFINITIONS

Radius of earth's orbit about sun	$= R_o$		m
Radius of earth	$= R_r$		m
Period of one orbit	$= T_o$	= 365	days
Period of one rotation	$= T_r$	= 24	hours
Velocity of center of earth in orbit	$= v_o$		m/s
Velocity at equator from rotation	$= v_r$		m/s
Latitude	$= \theta$		°
Velocity at latitude θ from rotation	$= v$		m/s
Maximum velocity at latitude θ	$= v_{max}$		m/s
Minimum velocity at latitude θ	$= v_{min}$		m/s

DATA

$R_o = (91.4 \text{ to } 94.6) \times 10^6$ mi — The World Almanac, World Almanac Co., 1990, pg. 246

$R_r = 3963$ mi — World Almanac, pg. 246

For New York

$\theta = 40°45'$ — World Almanac, pg 265

UNIT CONVERSIONS

$$R_o = \left(\frac{91.4+94.6}{2}\right) \times 10^6 \text{ mi} = 93 \times 10^6 \text{ mi}$$

$$= 93 \times 10^6 \text{ mi} \left(\frac{1609.3 \text{ m}}{1 \text{ mi}}\right) = 1.496 \times 10^{11} \text{ m}$$

$$R_r = 3963 \text{ mi} \left(\frac{1609.3 \text{ m}}{1 \text{ mi}}\right) = 6.378 \times 10^6 \text{ m}$$

$$T_o = 365 \text{ d} \left(\frac{24 \text{ h}}{1 \text{ d}}\right)\left(\frac{3600 \text{ s}}{1 \text{ h}}\right) = 3.154 \times 10^7 \text{ s}$$

$$T_r = 24 \text{ h} \left(\frac{3600 \text{ s}}{1 \text{ h}}\right) = 8.64 \times 10^4 \text{ s}$$

A. Find the orbital velocity

$$v_o = \frac{2\pi R_o}{T_o} = \frac{2\pi(1.496\text{x}10^{11}\text{m})}{3.154\text{x}10^{7}\text{ s}} = 2.980\text{x}10^{4}\text{ m/s}$$

The velocity of the center of the earth in its orbit about the sun is v_o = $2.980\text{x}10^4$ m/s

B. Find the rotational velocity at the equator

$$v_r = \frac{2\pi R_r}{T_r} = \frac{2\pi(6.378\text{x}10^{6}\text{m})}{8.64\text{x}10^{4}\text{ s}} = 4.64\text{x}10^{2}\text{ m/s}$$

The velocity of a person at the equator, caused by the rotation of the earth is v_r = 464 m/s.

To calculate the distance from the earth's axis to its surface at a latitude of θ

$$R = R_r\cos\theta$$

C. Calculate the velocity at a latitude of θ caused by rotation

$$v = \frac{2\pi R}{T_r} = \frac{2\pi R_r\cos\theta}{T_r}$$

At θ= 40°45

$$v = \frac{2\pi(6.378\text{x}10^{6}\text{ m})\ (\cos 40°45')}{8.64\text{x}10^{4}\text{ s}}$$

$$= \frac{2\pi(6.378\text{x}10^{6}\text{ m})\ (0.7576)}{8.64\text{x}10^{4}\text{ s}} = 3.51\text{x}10^{2}\text{ m/s}$$

The velocity of a person at a latitude of 40°45′, caused by the roatation of the earth is v = 351 m/s.

D. To find the maximum and minimum speeds

$$v_{max} = v_o + v = 2.980\text{x}10^4 + 3.51\text{x}10^2 = 3.015\text{x}10^{4}\text{ m/s}$$

The maximum speed of a person at a latitude of 40°45′ is v_{max} = $3.02\text{x}10^4$ m/s.

$$v_{min} = v_o - v = 2.980\text{x}10^4 - 3.51\text{x}10^2 = 2.945\text{x}10^{4}\text{ m/s}$$

The minimum speed of a person at a latitude of 40°45′ is v_{min} = $2.95\text{x}10^4$ m/s.

E. v_o and v are in the same direction when you are on the side of the earth away from the sun, that is, at midnight.

The maximum speed occurs at midnight.

Index